PRAISE FOR HOW TO C…

"A timely book that combines science with compelling narrative. Briggs gives us an expanding sense of possibility and hope with each page as she investigates what it means to be more connected, more human, and to feel more alive. I highly recommend this book!"
—Scott Barry Kaufman, PhD, host of *The Psychology Podcast* and author of *Transcend: The New Science of Self-Actualization*

"I have been waiting for this book. Saga Briggs offers us this satisfyingly scientific yet relationally centered emergence, a profound exploration of connection and direction for depth in inner work. As a psychedelic psychotherapist and, moreso, a human in a body who loves other humans in other bodies, Briggs's book is timely, expanding upon current somatic and psychedelic research towards our most integrated sense of self. If you are interested in, as Carl Jung put it, knowing thyself, this book is for you."
—Danielle M. Herrera, LMFT, psychedelic psychotherapist

"Having long felt a mismatch between my own experiences of guiding people through the psychedelic therapy journey, and the dominant discourse around mechanisms of action, it was a joy to read this book. Interoception (being able to really hear our inner healer speak to us through communication from our viscera, our deepest places) is very fully explored here and is therefore a must-read for those looking to deepen their understanding of psychedelic therapy. The author focuses on connectedness and embodiment, and the long, slow, deep journey of coming home to our wholeness, together, rather than brain resets and quick fixes that our individualistic culture craves and promotes. The book is a tender weaving between a personal story alongside diverse and fascinating findings of researchers. An important step in bringing more balance, integration, and nuance to the healing narratives of our time." —Dr. Rosalind Watts, founder of ACER Integration

"We live in a challenging era. Now, more than ever, there is no time to waste. Connection with ourselves and with the transient physical vehicles that carry around our conscious sense of self is of uttermost importance. Countless other writers and thinkers have given us these lessons for millennia. But in this book, Saga Briggs brings home this essential message in such a meaningful and contemporary manner that one cannot help but wake up, sit up, look within, and find solace and serenity in embracing ourselves and our bodies. Briggs teaches us to be with, not just in, our bodies. In doing so, wasting time becomes a delicious extension of possibilities. And just in time."
—Dr. Ben Sessa, psychiatrist, psychedelic researcher, and author of *The Psychedelic Renaissance: Reassessing the Role of Psychedelic Drugs in 21st Century Psychiatry and Society*

"Reconciling the sensory and embodied nature of experience with its cognitive and narrative dimensions is the next crucial task for the psychedelic renaissance. With a compelling synthesis of state-of-the-art science and first-person practice, Briggs highlights the key role of the body in psychedelic experience and mental health. If you would like to continue where Michael Pollan left off, this book is for you."
—Marco Aqil, founder of Amsterdam Psychedelic Research Association

"Interoception is a hot topic, with many open questions. Whilst scientists grapple with fundamental questions about the relevance of interoception for key psychological processes, for now it is important that books like Briggs's help the public to understand what interoception is and why it is important. Not only is it worth asking ourselves *how* am I feeling, it is also worth considering *what* am I feeling."
—Jennifer Murphy, PhD, Royal Holloway University of London

"*How to Change your Body* is an important book not only as psychedelic-assisted therapies gain momentum, but at this moment in the modern world in general. It speaks to a deeply feminine wisdom of the body, of connection, and human life that has for centuries been ignored or oppressed in society and in modern medicine. It presents a thorough scientific argument for the importance of interoception to mental and emotional well-being and to authentic, loving human connection."
—Dr. Devon Christie, certified MDMA and Relational-Somatic Therapist

"As a psychologist and clinical investigator in psychedelic clinical trials, I have had the privilege of witnessing the transformative power of these compounds. The profound healing potential of these substances has long been neglected by mainstream science, but recent advances are finally shedding light on their therapeutic benefits. This book brilliantly captures the essence of this re-emerging field, weaving together an impressive tapestry of science, personal insight, and the importance of the body in the development and manifestation of emotional health. This book challenges traditional notions of mental illness and empowers readers to reconsider their own relationships with their bodies and minds."
—Ingmar Gorman, cofounder and chief executive officer of Fluence

"Undoubtedly, readers of Briggs's cutting edge text on interoception, health, and psychedelics will find multiple synergies here: in insights and practices for both clinicians and those looking for self-healing. In addition to thoughtful interviews with practitioners, this book offers an innovative and beneficial toolkit of practices for enhancing and decolonizing our natural capacities for health and balanced interoception. Faulty interoception has been overlooked by mainstream mental health models and this book brings those missing pieces into clarity. Briggs contributes soundly to the tradition of utilizing psychedelics to discover natural mechanisms of transformation within our being. This book is a tour de force on the wisdom of the body."
—Janis Phelps, director of The Center for Psychedelic Therapies and Research, California Institute of Integral Studies

HOW TO CHANGE YOUR BODY

THE SCIENCE OF INTEROCEPTION & HEALING THROUGH CONNECTION TO YOURSELF & OTHERS

SAGA BRIGGS

FOREWORD BY JULIE HOLLAND

SYNERGETIC PRESS
SANTA FE • LONDON

Published by Synergetic Press
1 Bluebird Court, Santa Fe, New Mexico 87508
& 24 Old Gloucester St. London, WCIN 3AL, England

Library of Congress Control Number: 2023940332

ISBN 9781957869100 (paperback)
ISBN 9781957869117 (ebook)

Cover design by Amanda Müller
Cover image: *Homme de dos* by Pierre Appell
Interior design by Howie Severson and Amanda Müller
Typesetting by David Good Design
Managing Editor: Noelle Armstrong
Design and Production Manager: Amanda Müller
Production Editor: Allison Felus

Printed in the United States of America

For Collin Anderson

"You know the feeling when you and another are 'on the same page'? That's not what I'm talking about. That's the mind reporting back and saying that 'the other person and I believe or want the same thing.' Connection is body-based. It's a knowing in the body. Which means you need to know your body."

—Karen Dobkins, neuroscientist,
University of California, San Diego

Contents

Foreword by Julie Holland, MD

ONE OF THE MOST MEMORABLE EPISODES OF *THE SIMPSONS*, EARLY ON IN the series, involved a self-help guru giving a presentation in an auditorium full of the citizens of Springfield. Encouraging them to get in touch with their "inner child," the speaker used a young boy in the audience as an example of how to be in the world. He asked Bart Simpson up onto the stage and interviewed him about what drives him. Bart shrugged his shoulders and stated succinctly, "I just do what I feel like." The guru was delighted by this response and exhorted his audience to do likewise. Eventually, the whole audience is chanting, "Be like the boy! Be like the boy!"

This is interoception in a nutshell: the ability to monitor what we are feeling. Knowing you're upset is a form of interoception. So is recognizing your own hunger or arousal. This seems like a very basic concept—the awareness of inner bodily sensations. You'd think it'd be easy as pie. Of course you can do it, right? But can you truly feel at home in your body, safe and able to trust your own sensations?

This crucial book for our time, by the inimitable Saga Briggs, identifies a range of common diagnoses and "disorders" that are associated with decreased interoception. People who are depressed often have dysregulated interoception, paying too much attention to some things and not enough to others. This is also true with addiction. In truth, many of us have learned to ignore that quiet little voice telling us how we feel. We've adopted attentional strategies that veer away from making that intentional connection with ourselves, which is often connected to our suffering.

Briggs shows us how, if we become adept at interoception, we will benefit, in both body and soul. In particular, we will have a higher resilience to stressors, which can translate into lower levels of anxiety overall. If we could truly tap into that authenticity—of how we actually, microscopically feel—it would bring us both peace and power.

When I was a psychiatrist in training in New York City, I was obligated to enter into psychotherapy, not as the clinician, but as the client. I remember early on in the process, my therapist would hear of some childhood event, and she would inevitably ask, "And how did you feel at the time?" I would be at a loss for words, often answering, "I have no idea how I felt back then." Still, she persisted, "Well, how do you feel about it now?" I was surprised to hear myself eventually respond, "To be honest, I don't even know how I feel right now!"

That took some getting used to, the concept that even I, as a psychiatrist in training, devoting myself to the life of the mind, wasn't really in touch with my own feelings.

And that's why this book is so crucial for us all. I'm not alone in this alexithymia (the inability to recognize or describe one's emotions). The sad truth is that more of us are less in touch with our bodies, and how we feel in them, than ever before. Our phones and laptops pull us out of our kinesthetic awareness and into a virtual plane where we can leave our bodies behind. These electronic silos disconnect us from ourselves and from each other. But this important book makes clear that social contact and attunement to inner sensation can remedy such body disownership.

As much as we may assume alexithymia is something more men have because they are often socialized to suppress their emotions, women are also acculturated to distrust our own bodies, which are inseparable from our emotional lives. We are encouraged to ignore our hunger pangs and repress our natural drives for sensual pleasure and orgasm. Even though I grew up in the "Let It Be," crunchy-granola '70s, a time of letting our hair grow and taking our bras off, we were still out of touch with our own hunger for food, and our own lust, because it wasn't "ladylike" to feast fervently or feel horny.

Today women are still given overt and covert messages that we should be pleasing to others, making sure that everyone has enough before we take more. We're reminded to get a consensus from others on just about everything before coming to a decision about fulfilling our own desires. If you were a woman growing in a certain era, your whole existence was framed in self-denial. It's really no wonder that many women don't always know what they want. They're not taught to heed their own body's instructions. I would argue that this learned self-censorship has something to do with religious and patriarchal oppression. We are taught that it is a sin to be a glutton or a harlot. And of course, there is a strong element of patriarchy in this puritanism.

In order to deal with what's sometimes called "emotional eating," I had to read books on intuitive eating, relearning what hunger actually felt like. I had to become acquainted with my own sense of satiety in order to understand my cravings around food and to learn to satisfy them fully.

This disconnection from bodily signals—and the possibility of healing through attunement—is also relevant to addiction and childhood trauma. The addiction literature is full of references to interoceptive accuracy, the degree of detecting and identifying bodily signals. It may be that people who've gotten used to numbing themselves become less adept at locating their psychic pain. We also know that people who've been traumatized in their childhoods often habitually suppress any input from the side of themselves that split off as a result of this trauma.

And yet, some of the most traumatized patients I've ever worked with have been very clear about conveying their inner worlds. Years back, when I was making the rounds—visiting patients early in the morning in their rooms with a group of psychiatric residents—I asked my patient in front of the small group standing around his bed, "Does anything hurt you?"

"Yes," he stated after thinking for a moment. "My family doesn't come to visit me."

How can one not be moved by this thoughtful, personal response to a superficial question?

Which brings us to psychedelics. This book rightfully makes the claim that this group of psychoactive drugs can encourage deeper, richer responses to life's biggest questions: *Who am I? What do I want?* Psychedelic-assisted therapy can shake us out of our bodily ruts as well as our cognitive ones.

Firstly, there is the medicinal plant cannabis, which allows a heightened kinesthetic awareness (where your body is in space) as well as an enhanced emotional awareness. I find cannabis to be heart-opening, primarily, but it also helps me notice my posture and where I am holding tension. More than anything, it becomes easier for me to get in touch with what I'm really feeling and wanting.

Then there is the class of medicines known as the entactogens, which includes MDMA (Ecstasy or Molly); these allow a "touching within," meaning an enhanced ability to access inner feeling states. In this way, MDMA, 2-CB, and other research chemicals can help people improve their interoception, as well as communicating that information to others.

Additionally, a recently published study showed that people who had repeated access to ayahuasca ceremonies were able to positively affect not only the way they thought of themselves in the world (their narrative self) but also their sense of embodiment. Much like interoception itself, psychedelics have the potential to restore homeostatic balance. Just as when we notice we are hungry, we stop to feed ourselves, under the guidance of psychedelics, we notice our souls are depleted. When we are deprived of connection and meaning, and that mystical oneness shows itself, we stop to nourish ourselves.

Greater interoception can be accessed through psychedelics, time spent in nature, somatic practices, authentic human connection, and a range of other practices—something known to healers for thousands of years but only now being widely recognized in research and its literature. Through exploring these techniques, I hope this book will teach you the importance of finding a home in your body, resting in the peace and deep trust that can result from that singular connection of body and mind.

Harlem Valley, New York
March 3, 2023

Acknowledgments

I WOULD LIKE TO THANK THE FOLLOWING INDIVIDUALS AND ORGANIZATIONS, who made this book possible: Donna Foster and Carla Perry for writing mentorship during my "social critical period" of adolescence; Fred Canada, for timely connection and inspiration; Kit Kuksenok for countless enriching conversations on the body; John Zarr for expanding the social-embodied possibility space every time we meet; Hallie Price and Marissa Maislen for early read-throughs; Karen Dobkins and Andy Arnold for the early conversations that got me hooked on interoception, with special thanks to Andy for his thorough fact-checking; the MIND European Foundation for Psychedelic Science, for catalyzing a significant period of my professional and personal growth; Doug Reil and Noelle Armstrong for believing in the book and bringing it alive, with special thanks to Noelle for phenomenal editorial guidance; and the entire team at Synergetic Press for their commitment to making the world a more connected place.

Introduction

> *"One of the goals of taking psychedelics is to find a way to a certain state of mind without having to take psychedelics."*
> *—Rick Doblin, MAPS*

DURING LOCKDOWN IN MARCH 2020, AS THE PANDEMIC DIFFUSED INTO life in Berlin, I had an upsetting but revelatory dream: I was a blank canvas, and other people were the paint that coated me, enlivened me, brought me into being. Without others, I was nonexistent, or at least incapable of seeing myself. Weary of keeping my body alive in such a state of crisis, my brain pivoted out of uncertainty and woke me up.

Oddly enough, in the days of research following the dream, I found that scientists, philosophers, and clinicians were starting to ground this idea in their work: the quality of our connection to ourselves begins with our connection to others—not in a good conversation over coffee, per se, but in the felt experience of "being seen" by another—and this quality is determined very early in our lives.

At the University of Jyväskylä in Finland, Joona Taipale, PhD, a senior lecturer in the Department of Social Sciences and Philosophy, believes individuality is a developmental process. We are not born with a sense of separateness from others, he says; rather, it's a gradual process of recognizing our otherness, which develops during infancy.

In developmental psychology, the prevailing notion is that the other—the "object"—is assumed to be differentiated from the self at the time of birth. But Taipale challenges this idea.

"In the psychoanalytic tradition, this picture is turned upside down," he writes in his 2017 paper "The Pain of Granting Otherness: Interoception and the Differentiation of the Object." "Instead of taking the separateness of others as a point of departure, it argues that the capacity to grant the 'otherness' of others is preceded by a long and complex development—a painful process in and through which the object is gradually placed outside the area of the subject's omnipotent control."

For example, as infants, there's no reason to assume our mother's breast (or our mother, for that matter) isn't part of us, until we develop a sense of agency. Over time, we come to realize this isn't the case, as we learn the breast (and mother) isn't always immediately there when we want it, but we begin life under the illusion that it is.

"From the point of view of an external observer, the infant and the mother are, of course, two different and distinct entities," he says, "but the situation looks very different if considered from the point of view of the infant."

Taipale's contention? We are born connected.

A team of French neuroscientists takes this notion one step further.

"While cognitive psychologists have long tended to consider mental functions—including self-awareness—as isolated entities, recent arguments pointed out the fundamental social nature of self-awareness," writes Nesrine Hazem, PhD, who conducts research for the Social and Affective Neuroscience Laboratory at ICM in Paris. "By experiencing the self as the object of another's attention, infants may develop an initial representation of self and others as psychological entities. Rather than suppressing self-experience in adulthood, social contact would then give rise to the experience of being a cognitive-affective agent. Thus, a fundamental property of social contact throughout the life span would be to enhance self-awareness."

Her team's own work substantiates the link between not only social contact and self-awareness, but the fundamentally body-based nature of both. "We showed that hearing one's own name spoken by another person and being touched by another person both increased bodily self-awareness, just as perceiving direct gaze does. This provides the first empirical demonstration that social contact irrespective of sensory modality elicits bodily self-awareness. In doing so, it supports the notion of the social nature of the self—i.e., that human self-awareness emerges in interpersonal contacts."

Why does any of this matter? Because nearly every mental illness under the sun—anxiety, depression, addiction, schizophrenia, eating disorders, posttraumatic stress disorder (PTSD)—is associated with impaired bodily awareness. Which, if the social self-awareness theory is correct, suggests they are interpersonal disorders too.

Psychedelic science, a rapidly growing field showing great promise in advancing human health, seems to be revealing a common thread throughout these illnesses: relational disconnection. Not just to ourselves, our bodies, and others—but also to the very fragile threads tying these three things together.

One of the main drivers behind the therapeutic effects of psychedelics is now thought to be a change in one's experience of social relationships. As the research is beginning to suggest, that means a change in one's experience of oneself and one's body. These things can't be separated.

But it wasn't until recently that relationships became a core focus of psychedelic science. Around the time Michael Pollan's book *How to Change Your Mind* came out in 2018, researchers believed the hallmark of a transformative psychedelic experience was "ego dissolution" (or "ego death") through a quieting of the default mode network, a region of the brain thought to drive self-referential thinking. Things have evolved since 2018.

"It's not so much the self-loss but rather the connectedness that seems to be driving the well-being," says David Yaden, PhD, who studies transcendent experiences at the Johns Hopkins Center for Psychedelic and Consciousness Research, in an interview with humanistic psychologist Scott Barry Kaufman. He concedes that there's a correlation between self-loss and connectedness, but his studies show that "only connection seems to be correlated with beneficial outcomes related to well-being. I think a lot of researchers are maybe barking up the wrong tree in terms of ego dissolution, and this feeling of social connectedness is where the real action is at. My prediction is that these processes related to social connectedness and attachment will end up being the real mediator in at least some of the therapeutic benefits."

If social awareness, self-awareness, and bodily awareness are inextricably linked, this implies that one important reason psychedelics work to treat a wide range of afflictions is because they have the capacity to change one's experience of bodily feelings, not just one's thoughts.

In Pollan's book, one depressed patient who had taken psilocybin, the psychoactive compound in magic mushrooms, described his experience as follows: "Once I was in this state, it was beautiful—a feeling of deep contentment. I had this overwhelming feeling—it wasn't even a thought—that everything and everyone needs to be approached with love, including myself."

Although the effect lasted a few weeks, it eventually faded in the following months and his depression returned. One year later he wrote, "The insights I gained during the trial have never left and will never leave me. But they now feel more like ideas."

This notion hasn't escaped more recent psychedelic studies. The Watts Connectedness Scale, for example, provides the first scientific model for measuring feelings of connectedness during psychedelic experiences. Dr. Rosalind Watts and her research team published a paper in 2022 describing their initial findings on the relationship between psilocybin and connectedness in participants with depression. As you might expect, study participants reported feeling more connected to themselves, other people, and the greater world after ingesting psilocybin. Interestingly, few people reported feeling more connected in a singular domain without experiencing greater connectedness in the others.

"Based on the findings of this study, we now predict, for example, that an individual reporting feeling connected to a sense of meaning and purpose as well as their body and emotions would also report feeling connected to other people; or that someone describing feeling connected to nature would also report feeling connected to humanity at large and their own emotions," they write. "Overall, high scores on one domain (self/others/world) suggest high scores on others, just as low scores in one aspect suggest a disconnection across multiple domains."

To me, this begs the question: is it that psilocybin "enhances" connectedness by making it feel multidimensional, or that connectedness itself is already multidimensional, and psilocybin reveals it to be so?

One clue, perhaps, is in the way study participants tend to describe their experiences of self-connection: "I have felt connected to my body"; "I have felt connected to my senses"; "I have felt connected to a range of emotions."

"Previous conceptualisations of 'self-connection' have been largely cognitive, emotional, and behavioral, and have not included embodied/somatic aspects," Watts and her team report. "Self-connectedness in psychedelic therapy tends to be described as connectedness to the senses, the body, and emotions."

When I consider how I've managed my own illnesses—substance abuse and anxiety—I notice that the things that worked for me functioned in a very similar way to how MDMA, a psychedelic-type drug known to enhance both bodily awareness and sociality, may work, without MDMA being part of the picture. It was a resocialization process, and it happened through my body.

Is it the bodily experience that we should be honoring and paying more attention to, with or without psychedelic drugs, when it comes to mental health treatment?

For the past three years, a large part of my journalistic efforts have been aimed at answering this question through the lens of interoception, the process of sensing the body from within. Increased capacity for interoception—considered by some to be the "eighth sense"—relieves mental illness and promotes well-being by helping us connect to ourselves and others, and plays an underacknowledged role in a wide range of wellness practices and treatments, from yoga to psychedelic therapy. What follows is a summary of what I've learned so far, from the scientific literature, from researchers and practitioners around the world, and from my own personal science and reflection. Much like psychedelic research, the research on interoception is just arriving, but that doesn't mean we can't use our imaginations, be right about the things we imagine, and start putting them into practice now.

This is a book about connection. It is also an argument for reframing mental health as relational health. Because the number one thing that soothes the nervous system, and does not damage it over time—that makes the body feel safe, allowing the mind to follow—is authentic human connection.

I. PREDICTING THE BODY

Predicting the Body

For six years, I drank to connect. It began in 2011 in Ohio, like the flip of an epigenetic switch, taking a shot of rum from the kitchen cupboard before going to meet some college friends I wanted to impress. Five years later, I found myself knocking back Negronis on a Monday night in Seattle until I couldn't stand up. Once I scaled it back, my withdrawal symptoms lasted a full year. It took me two more to stop drinking completely.

When we talk about using substances to "numb," we're suggesting that we don't want to feel. But humans never stop feeling. As the neuroscientist Antonio Damasio said, "Humans are not thinking machines; humans are feeling machines that think." I used alcohol to control the way I felt, and to reduce uncertainty around how I would feel. I didn't drink to numb. I drank to predict.

PREDICTIVE CODING

One of the most widely accepted accounts of human brain function is the predictive coding theory, which argues that the brain evolved to generate a reliable model of the world around us and eliminate false guesses as efficiently as possible. According to this theory, the brain creates models for our perception, concepts, and feelings based on experience, and anything falling outside these models results in a "prediction error," which is then either updated by new information (adapting to the temperature of a hot bath) or by prior beliefs (the feeling of having a real limb where one is missing, as in phantom limb syndrome). Although it's meant to serve an adaptive function, too much prediction in the wrong direction can trap us into destructive belief systems and habits, including substance abuse.

I used to know exactly how confident I would be on a date after two gin and tonics. I knew how thick to slice a lime to stuff the pulpy wedge into the neck of a Pacifico, and how the first sip meets your lime-slicked lips and jazzes on the tongue. I knew the protective warmth of Yellow Tail shiraz, sails lifting the solar plexus, expanding the space between the ribs. I knew the proud feeling of being the cool girl who orders a White Russian or a whiskey neat.

When I went out with my boyfriend and our friends in Portland, Oregon, back in 2012–2013, I would usually start with cocktails or a shot, then switch to beer. Whatever chemistry occurred in my brain as a result of that order of things seemed to improve my memory, focus, and attention in a way that made socializing easier. Normally, for example, I might be trying to tell a story and be hypervigilant about other people's perceptions of me while I was telling it, which takes mental resources away from the actual telling of the story. But liquor made it easier to eliminate that hypervigilance and focus on myself, which led to better storytelling. I came to associate a Fireball and a High Life with social performance, and my prediction rarely failed me. I also knew that if I started the night with beer or wine, my thinking became muddled and my brain fuzzy, which made it harder to hold a conversation, so I avoided that.

Predictive coding affects the entire nervous system (e.g., "knowing something in your heart" has a real physiological basis) and is closely tied to a process called interoception, which is our sense of what's going on inside our bodies. Any jitters I had before that first drink are an example of the elegant interplay between interoception and predictive coding: my body recognized an environment (bar, social gathering) where it could predict the conditions under which I would be nervous or relaxed, and if the conditions for relaxation weren't yet available (read: prediction error), I'd feel nervous. Once I consciously picked up on that nervousness, the predictive coding kicked into full force and I couldn't relax until I resolved the prediction error with a shot and a beer back. Alcohol dependence became a rigid predictive mold encasing my nervous system.

"In the most general terms, interoceptive perceptions—that is, what is experienced—derive from the brain's best guess about the causes of events in the body, with incoming sensory inputs keeping those guesses in check," says Lisa Feldman Barrett, PhD, who pioneered the EPIC (Embodied Predictive Interoception Coding) model of cognition. "Not only has your past sensory experience reached forward to create your present experience, but how your body feels now will again project forward to influence what you will feel in the future. It is an elegantly orchestrated self-fulfilling prophecy, embodied within the architecture of the nervous system."

In other words, not only does information flow from our senses to our higher faculties to be processed so we can understand and interact with the world, but those higher faculties also often "predict" the input from our environment, thereby influencing our perception of the world before we actually get a chance to sense it.

"You experience, in some sense, the world that you expect to experience," says Andy Clark, DPhil, a cognitive scientist at the University of Edinburgh in Scotland. "All experience is controlled hallucination."

Again, the reason for this controlled hallucination is efficiency: think of the way a computer stores video files, which contain enough redundancy from one frame to the next that it's more efficient to encode the differences between adjacent frames and then work backward to interpret the entire video than to encode every pixel in every image when compressing the data. Memory is thought to work the same way, eliminating nuance and preserving the gist of the experience.

But when it comes to circumventing this self-fulfilling prophecy, and not only creating behavior change but rewiring your own predictions, the more energetically efficient solution may be to extract the data and consider each pixel. And when that happens, the body—not the mind—is the first player up to bat.

"Recently, there has been much interest in the concept of predictive coding in interoception," neuroscientists Karen Dobkins, PhD, and Andy Arnold, PhD, write in a theoretical paper on the topic. "Here, the notion is that, as part of homeostatic processes, the interoceptive system tries to minimize errors between what is predicted versus actual, since errors are energetically expensive." While they agree that prediction errors are energetically expensive, they also note that reinforcing one's predictions can itself be expensive, especially in the social domain. "A concrete example of this might be 'getting worked up' about an upcoming presumed-difficult conversation. This can be exhausting, not to mention the fact that bracing for a difficult conversation can lead to a negative self-fulfilling prophecy. We suggest that the most energetically efficient solution is to acknowledge one's priors and yet go into a social interaction with a 'beginner's mind,' staying present to, and trusting, one's interoceptive signals as the situation unfolds."

To sit with uncertainty—the feeling, not the concept—is the first step toward knowing.

INTEROCEPTION

One rainy day in Portland, in the fall of 2016, I was writing and listening to Bob Dylan's *Blood on the Tracks* when I noticed my fingers trembling as I held them over my keyboard. I knew what was happening. I'd been experimenting with drinking less in the previous few months, and this was my body's reaction to it. The strange thing was that I'd felt fine on the inside, at least most of the time, for the entire five years prior to cutting back. But it struck me in that moment that I must not have been fine, despite thinking I was fine, and that somehow I must have missed the part where my body warned me things were getting bad.

Instead of noticing two main body states—less relaxed before an expected drink and more relaxed after an expected drink—I began having no frame around my feelings. One night I helped my brother-in-law set up appetizers in his kitchen as guests arrived for dinner, and the anxiety was like the flash of a camera inside me, a photonic fear, my body sublimating on the spot. I worried about my body most of all—how to arrange it in space around others, where to be, how to stand, how to occupy time and space in a suitable embodied rhythm. Speaking was a physical act, too: would I try to be connected to the words coming out of my mouth, or would it be easier to stay disembodied and follow a script? The more I thought about these things, the less grounded in my body I was and the less of a chance there was for my authentic self to show up at the party.

In 2017, I went to Paris to stay with a French boyfriend I'd met while traveling. In some ways I could not summon the life force to participate in normal things. I couldn't rent bikes with him because the whole thing seemed to require too much energy. If we weren't drinking, I'd just want to go back to his flat and lie in bed and watch a movie while the sun shone outside and tourists scaled the Eiffel

> Tower. We were doing the non-touristy thing, I told myself, just living the way Parisians live. My hand shook under the blanket, completely out of control, as we snuggled up around his laptop screen. Alongside these withdrawals, I started noticing these moments of love for him that filled me completely. It was everywhere in my body at once. My chest, my arms, parts of my body I couldn't name. It seemed to glow without a point of origin.
>
> It was as if I'd opened a floodgate in the previous six months and all sensations—pleasant and unpleasant, exaggerated and exact—were rushing through at once. While some of the feelings were transformative in a positive way, many were debilitating. In fact, it was quickly becoming clear that I could not love a person fully while my body lived in fear. I wanted to know: how could I regain control over my situation, how could I inhabit a different reality, how could I change my body?

All of these experiences fall under the category of interoception. As I said before, interoception is the process of sensing the body from within. It's how we know when we're thirsty, hungry, horny, joyful, anxious, tired, angry, sick, elated. A lesser known sense compared to the main five (which are collectively called "exteroception"), interoception is like mindfulness for the body—it's what allows us to be aware of our own well-being. And it's unique for everyone, although it's generalizable enough that we can often understand each other's descriptions of interoceptive sensations.

You may have noticed that emotional and physical feelings are part of the same list. That's because the interoceptive process that allows us to sense the state of our internal organs also allows us to experience and regulate emotion. Although it wasn't always believed that emotions arise from anywhere but the mind, neuroscientists have found that the same part of the brain—the insular cortex (or insula)—is a primary hub for body sensing and emotion processing. We tend to think that signals for emotions like joy or anger have nothing to do with signals for body processes like hunger and thirst, but this is a mistake. They are part of the same interoceptive system. Body and emotion are inseparable.

Along these lines, Dobkins and Arnold argue that loneliness is a "social hunger signal," managed by the same process as hunger for food. In their view, a healthy social existence is not defined by the number of connections you have, but by listening closely to what's going on inside your body.

"At least for physiological needs," they say, "we (and all animals) have evolved mechanisms to 1) sense current amounts of resources as well as 2) motivate behavior to acquire needed resources. For example, 'hunger' is a sensation that signals a lack of food, inciting us to acquire that resource, else we would die of starvation. The mechanism for sensing hunger is *interoception*—a system for sensing internal physiological states of the body (e.g., hunger, thirst, temperature) as a means of regulating and maintaining homeostatic conditions."

Similar to physical hunger, they say, social hunger should serve to motivate social (re)connection. They propose that "interoception plays an important role in indicating whether one feels lonely or socially connected—i.e., adequate interoception may buffer one from loneliness."

But you have to pay attention to your body to recognize the signal.

"One may ignore (or be insensitive to) one's feelings of loneliness, similar to ignoring physical hunger pains, and therefore not be motivated to (re) connect with others," they write. Or you might recognize the signal but avoid responding to it socially because of a negative past experience: "One may actively avoid the social world because past social interactions have been low quality and therefore, unrewarding." Over time, this could also lead us to ignore the signal.

The point to remember is that bodily signals often precede emotion, and it is specifically our awareness and interpretation of those signals that allow the construct of emotion to arise at all.

One of the first psychologists to suggest that body states precede emotions was William James. James believed our heart pounds not because we are anxious, but rather that we're anxious because our heart pounds. Certainly, believing we're anxious may cause our heart to pound even harder, but the reason our heart starts pounding initially is for a fairly neutral reason: it's alerting us to something important, something that most likely requires our attention and energy. It's our subjective interpretation of the pounding heart as anxiety, fear, or panic that induces the unpleasant emotional state that follows. The bottom line is that we can learn to interpret—and respond to—the signal in healthier ways.

You can already start to see how faulty interoception might lead to sustained illness.

One of the world's foremost authorities on interoception is Sarah Garfinkel, PhD, a clinical affective neuroscientist at University College London. She subscribes to James's notion of emotions arising from body states, and thinks interoception is behind it. Much of her work focuses on heartbeat perception, which is closely related to emotion in the brain. What often happens in the case of mental illness, she believes, is that people become disconnected from truly understanding what's happening inside their bodies.

"It was actually PTSD that made me first get interested in the heart and interoception," she says.

She was interested in the mechanisms in the brain that give rise to persistent fear memories: why do individuals with PTSD—a condition marked by past experience or exposure to a traumatic event—feel frightened in the present, when there is no longer anything to fear? Garfinkel figured there must be something going on internally to drive the fear response. Not only do PTSD patients have hyper-arousal symptoms including a fast-beating heart, they also show hyperactivity in brain areas related to emotion such as the amygdala and insula—key structures that are dynamically related to the heart. She thinks there must be a subliminal fear signal that appears more quickly than these people can be consciously aware of, making their hearts beat rapidly even though "rationally speaking" they might not report any threats in their environment.

The clinical literature has already established that interoception, like predictive coding, plays a major role in most mental illnesses, including anxiety, depression, PTSD, schizophrenia, addiction, panic disorders, eating disorders, and OCD.

"Understanding interoception and how our body interfaces with our brain and mind is going to open massive avenues and doors and windows and insights into the nature of different psychiatric conditions," Garfinkel says. "What we know at the moment is only just starting."

But before we get into the potential of interoception for managing illness, it's worth turning our gaze backward and considering how it develops in the first place.

THE BODILY SELF

> *"We look at the world once, in childhood. The rest is memory."*
> *—Louise Glück, American poet*

> I grew up in the majestic wilderness of the Oregon coast, rooted deeply in a family of caring souls with unresolved traumas. My parents were both medical professionals and artists—my mom a nurse manager and a painter, my dad a respiratory therapist and musician. They spent an enormous amount of effort tending two acres of land in a river valley, year after year, with groves of plum trees and redwoods, seasonal vegetable patches, and perennial flower gardens. Some nights, they'd fall into screaming matches after I'd gone to bed—a sound bath of slamming doors and ultimatums. In the morning they'd be back outside, planting willow trees along the river's edge to stabilize the bank.
>
> A few years ago, my mom found a bookmark I'd made her when I was five or six. Underneath a drawing of a tulip, I'd written, "A mother is a beautiful flower. Sometimes it droops. That's why I am made of so much water."
>
> Recently, I met a renowned psychotherapist at a psychedelics conference who, noticing I'd been waiting to speak with him while someone else delivered a monologue into his ear, reached out very gently and touched the name tag on my lanyard and said, "What about you?"
>
> The waters stirred inside me.

At Utah Valley University, the pioneering work of Kristina Oldroyd, PhD, suggests that early social experiences significantly impact areas of the brain responsible for interoception by influencing the development of the bodily self. Oldroyd's research team has found that insensitive caregiving—which includes responding inconsistently to a child's needs or rejecting distress altogether—can impair a child's ability to form accurate representations of bodily sensations. For example, when a child who is learning to walk falls down and feels physical pain, a sensitive response from a parent might be, "That must have hurt," whereas an insensitive response would be, "You're fine, that didn't hurt, get back up." For the child to become comfortable detecting, acknowledging, and expressing bodily cues, the parent must notice what the child is experiencing, draw joint attention to it, and label it:

> To the extent that caregivers recognize, honor, and respect their children's bodily experiences, the child will develop more accurate interoception. To the extent that a child's bodily experiences are denied, devalued, ignored, or punished by parents, the child will find ways to avoid feeling them, and develop a distorted sense of interoception.

In her 2019 study, Oldroyd and her team measured how comfortable mothers were with the negative emotions of their children, aged 8 to 17, and found that level to be highly correlated with how connected their children were to their bodies. "The more moms were able to deal with negative emotion and to help kids cope with negative emotion, the better their kids were at recognizing and responding to their own internal feelings of anger or emotional distress."

Some of the measures on the acceptance/rejection of negative emotion scale were "When children cry they're being manipulative"; "I want my child to experience negative emotion"; and "I'm afraid when my child displays negative emotion."

Say a child is about to perform a dance recital and has an increased heart rate and her palms are sweaty. Her parent could either say, "That's really normal; your body's preparing for a challenge. This is your body's way of telling you you're prepared; you're going to be great." Or they could say, "Oh my gosh, you're having a panic attack; let's pull out a Xanax."

Other researchers are adding to Oldroyd's findings, showing that mothers' interoceptive knowledge about their own emotions is associated with children's social affective skills (emotion regulation, social initiative, cooperation, self-control), even after controlling for child gender and ethnicity, family income, maternal stress, and the above maternal socialization factors. Findings suggest that mothers' interoceptive knowledge "may provide an additional, unique pathway by which children acquire social affective competence."

Oldroyd maintains that the way we learn to regulate physical pain is no different from the way we learn to regulate emotional pain—in both cases, we are socialized through our bodily experience. Neuroscientific studies support her theory, showing that children's attachment styles are linked to changes in the interoceptive network. If the bodily self remains unchanged throughout those children's adult lives, when relationships become more complex and social-emotional regulation increasingly important, Oldroyd believes it is poor interoception itself that may lead to disorders like anxiety, depression, and addiction.

It's important to note that there are three facets of interoception—interoceptive accuracy, interoceptive sensibility, and interoceptive awareness—that all play different roles in childhood development and adult well-being. I spoke to Kristina Oldroyd in the spring of 2020 about these different shades of interoception and how they develop in the early stages of life.

SB: What is interoceptive accuracy?

KO: Interoceptive accuracy (IAcc) refers to an individual's ability to detect and track internal bodily cues. IAcc is an objective empirical measure of behavioral performance, distinct from a person's subjective feelings about the body.

SB: What is interoceptive sensibility?

KO: Interoceptive sensibility (IS) refers to an individual's style of interpreting their bodily sensations. The measure of interoceptive sensibility is meant to capture people's judgments and beliefs regarding their bodily cues and thus allow researchers to assess how a person *thinks* about what they *feel*.

SB: What is interoceptive awareness?

KO: The third component of interoception is interoceptive awareness (IAw) and is defined as a metacognitive measure that quantifies individuals' explicit knowledge of and confidence in their interoceptive accuracy. In other words, interoceptive awareness is operationalized as a degree of error, the quantifiable difference between (i) self-reported judgment about one's interoceptive accuracy and (ii) objectively assessed performance on tests of interoceptive accuracy. The former is assessed by questionnaire, the latter with the heartbeat-tracking task.

A high level of interoceptive awareness reflects a person's ability to consciously know whether or not they are making good or bad interoceptive decisions based on their perceived IAcc.

SB: What role do early social experiences play in developing interoceptive accuracy?

KO: Early social experiences have been found to affect the anatomy of the insula, with children who are classified as having an anxious or an avoidant attachment style demonstrating markedly lower insular volume and smaller surface area than control groups. Additionally, attachment-related processes have been shown to affect electrical activation in the insula such that people with an avoidant attachment style showed decreased insular activation in response to stimuli than do securely attached individuals. These insular changes suggest that individuals with an anxious or avoidant attachment style may demonstrate differing levels of interoceptive accuracy as compared to people with a secure attachment.

SB: What role do early social experiences play in developing interoceptive sensibility?

KO: In young adults with an avoidant attachment style, the anterior cingulate cortex (ACC) has been shown to fail to fully integrate with the insula. This neurological pattern has been correlated with a blunted emotional affect and hypo-activating strategies in the face of distress. In other words, these individuals do not respond as strongly to emotional stimuli as expected. Given this, I would expect that in individuals with an avoidant attachment style, where the ACC/insula connection is impaired, interoceptive ability would also be impaired. This means that the early social experiences that lead people to develop an avoidant attachment style may also lead them to develop the propensity to be slower to recognize bodily cues and slower to interpret bodily cues as important once felt.

SB: How can a condition like anxiety arise from poor interoception?

KO: While early theories suggested that heightened sensitivity to bodily cues was at the root of anxiety, more current work suggests that anxiety symptoms arise from discrepancies between a person's actual and expected bodily state. Thus, it is the bias to interpret bodily signals in a negative manner (interoceptive sensibility) rather than the noticing of bodily signals (interoceptive accuracy) that contributes to both the cognitive (e.g., worry) and behavioral (e.g., avoidance) symptoms of anxiety.

In a later chapter, interoception researcher Andy Arnold will tell me how, somewhat counterintuitively, better interoception leads to a more distinct boundary between yourself and others. If you are in touch with and honoring your own bodily experience, you don't let it get hijacked by other people's emotions or opinions. The feeling of greater connection to people under the influence of psychedelics may not be at odds with this. If psychedelics can enhance our interoceptive awareness and feelings of closeness to others, perhaps it's not because the boundaries between you and others dissolve but because they become more clearly established. You sense yourself better; therefore, you have a better capacity to feel where another person fits into that visceral world, how to engage and mix energies with them. A thing can be greater than the sum of its parts, but one prerequisite to that greatness is the individual nature and inherent separateness of those parts. You are your own point of reference in the relational world.

THE POSSIBILITY SPACE

> *"The possibilities that exist between two people, or among a group of people, are a kind of alchemy. They are the most interesting thing in life." —Adrienne Rich, American poet*

In January 2018, I moved to Berlin after spending four years as a digital nomad: two years with the Portland partner and two years on my own. Something about Berlin felt like home when I first visited, but I also figured that moving to a city full of strangers, living in a flat full of strangers, and meeting as many people as possible, as soberly as possible, might help me rewire my relational existence to some degree.

During that first year, psychedelic research entered mainstream awareness with the publishing of Michael Pollan's book and rumors of the opening of the Johns Hopkins Center for Psychedelic and Consciousness Research. Although I'd taken mushrooms with friends in college, I truly became interested in psychedelics after reading an article in *Quartz* on psilocybin and ego dissolution. The day before I stumbled upon the article, I'd been thinking about how recovery seemed to require a reckoning with the ego-driven roots of addiction, particularly in social settings. Although my focus would later shift from "ego" to "attachment style," my research led me to the literature

> on psychedelics and mental health, and of course I began thinking about anxiety, addiction, and other issues in a different light. Before long, I'd written and published an article for the Berlin-based MIND Foundation for Psychedelic Science on predictive coding and its antidote: possibility.

Psychedelic science offers some of the best evidence for predictive coding by showing us what happens when it is compromised. Instead of simply increasing brain activity, substances like psilocybin and LSD loosen the predictive filter that's normally in place, allowing us to perceive other possibilities—for example, being arrested by the incredibly vibrant detail of a daffodil rather than passing it off as just another flower in the garden.

A research team at the University of Minnesota explains: "Psychedelic drugs perturb universal brain processes that normally serve to constrain neural systems central to perception, emotion, cognition, and sense of self."

If you think about it this way, the cognitive dissonance you might experience when you reach up to adjust your glasses only to remember you put in your contact lenses a few hours ago could be considered a kind of "trip": your brain generates a prediction error the same way it does when you suddenly witness the surprising vibrancy of the flower. It's all about the contrast between what's expected and what's experienced.

So why does any of this matter? What does it change? Neuroscientists usually talk about predictive coding in terms of sensory perception, but it also applies to relationships with ourselves and others. Staring at daffodils all day may not be efficient for survival, but it's not a meaningless activity either. The underlying perceptual setting—an openness to possibilities—is the same setting that nudges us out of depressive spells, lets us see things from another point of view, and fuels deep thinking and creativity.

In terms of social life, most of us follow unspoken scripts of relational behavior, from the activities we pursue together to the language we use to express our thoughts and feelings to others. Often, these scripts limit our ability to explore the possibilities poet Adrienne Rich was writing about. We go out to eat and drink; we say "I love you" and "I'm sorry." Languages and cultures differ in the scripts they offer, but no language or culture is free of these limitations.

Language itself is a gargantuan limitation, and feeds the predictive part of the brain. The more we say and hear "I love you," the more we expect to say and hear it, and the more we lose curiosity about the experience of love itself. Another way to put it might be that the more we talk through our experience, the less we truly experience it. Categories and conclusions help us communicate our experience at the expense of reality, or at least at the expense of precision and nuance. The more we go to restaurants, bars, parks, concerts, and football games, the more it seems these are the only activities people can do together. Not that there's anything wrong with these activities—just that doing them means *not doing* something else that, if imagined and pursued, might make life even richer and more enjoyable. Relational scripts help us make sense of our relationships for the sake of drawing categories and conclusions more than they allow us to explore them.

We also learn to predict people, including ourselves. I'm guilty of ending relationships with people I found too predictable. Another possibility might have been to take a closer look. I've also been known, on occasion, to trap myself in self-defeating mental narratives that leave no room for growth. On my good days, I try to cultivate awareness of the existence of other possibilities, even if I can't imagine them yet. There is always another reality, like an extra chair at the same table, right next to the one you're sitting in. Your chair might be wobbly or well grounded, but I think that metaphorical proximity—right next to you—is important. Relational possibilities fit within this metaphor too. The hard part is lifting yourself under the enormous weight of prediction and moving to the other chair.

Seeing other possibilities is extremely useful when it comes to mental health. Neuroscientists now have reason to believe that psychedelics work to treat conditions like depression and anxiety by influencing the mechanics of predictive coding in this way. One typical symptom of depression, for instance, is rumination, an inescapable loop of thoughts revolving around low self-efficacy or inferiority. Neuroscientists think this loop is interrupted when you introduce a drug like psilocybin or ketamine to the system.

"To my mind," says Philip Corlett, PhD, associate professor of psychiatry at Yale University, "these drugs expand the possibility space—the number of models that could be learned over and about—and so they bump people out of their depression/anxiety rut."

Crucially, we are not just talking about a mental rut. It's a body rut too.

Psychedelic research teams around the world are starting to corroborate the view that how one feels about one's life has more to do with the body than we might suppose. "The consistent effects on phenomenal alterations of the bodily self, coupled with the significant alleviation of clinical symptoms following ingestion of psychedelic substances, allude to a potential role of embodied cognition in these altered experiences," writes a team at the University of Zurich in Switzerland.

Meanwhile, at the University of Leiden in the Netherlands: "Psilocybin . . . could allow for an influx of exteroceptive and interoceptive information. Such influx may lead to increased interoceptive awareness, which has been associated with awareness and regulation of emotional states. A neurocognitive mechanism which may underlie the effects of psilocybin on emotion processing and interoceptive awareness can be found in the predictive processing framework."

Although a region of the brain called the thalamus is thought to mediate this influx of information, the insular cortex—which, as we've seen, affects social cognition—likely plays a key role as well. In a 2017 Multidisciplinary Association for Psychedelic Studies (MAPS) study, following research correlating heightened insula activation with PTSD and social anxiety, Robin Carhart-Harris, PhD, and his colleagues concluded

> The present findings highlight insular disintegration (i.e., compromised salience network membership) as a neurobiological signature of the MDMA experience and relate this brain effect to trait anxiety and acutely altered bodily sensations—both of which are known to be associated with insular functioning. Our study represents the first synthesis of these lines of research, demonstrating decreased insula/salience network functional connectivity under MDMA and linking this to baseline levels of trait anxiety and changes in interoception.

When I asked Kristina Oldroyd to comment on this work, and how MDMA might impact interoception, she said it most likely affects interoceptive sensibility: "The neural impacts of MDMA could potentially translate to improved abilities to process body-state relevant information, thereby impacting how an individual engages in motivated behavior to regulate their internal feelings rather than in different interoceptive feelings."

Psychedelic drugs—which we often take to feel more connected not only to ourselves but also to others—can bump us out of the body rut.

I did not use psychedelic drugs to work through my anxiety and addiction. But I did practice expanding my own embodied possibility space by trusting my body in the presence of others.

In 2019 I chose sobriety, as opposed to moderation, because of predictive coding, and because of interoception (though I didn't know what it was then). It wasn't that I didn't trust myself around alcohol, at least not in the conventional way of speaking about addiction. I'd already gone several years drinking in moderation, and I didn't get cravings or have any issues limiting intake. I trusted my mind. But there's a difference between the mind and the brain, and I didn't trust my brain as much. I knew that if I gave my nervous system the option of relying on anything but my own body to relax in a social situation, it would not allow me to reach that deep level of body trusting I needed to heal. My brain needed to know that it could not count on—or predict—alcohol under any circumstances. Otherwise, the nerves would perk up now and then—like little question marks—wondering whether they could make an exception in this case and depend on a substance. So I gave my brain one option, which also happened to be a new possibility: trust your body.

Over time, by gently falling into that visceral landscape again and again, I seemed to override the programming I'd installed with alcohol and replaced it with a calm, relaxed state grounded in trusting my body. It's not that I never suffer anymore, but my suffering is significantly reduced to the point of being negligible compared to what it was before. Along the way, I wasn't just healing my alcoholic-anxious self. I was healing the self that had chosen to drink in the first place.

All of this to say, the label "mental illness" no longer suffices. Not only is the mind inseparable from the body, but the body is inseparable from social relationships. Human afflictions are biopsychosocial, closely tied to emotion states arising from relational bodies. This is something long known by clinicians and scientists, but still the notion of a "chemical imbalance" dominates narratives of illness, just as the meaning of thought patterns tends to outweigh the significance of bodily signals in psychiatric settings. The biopsychosocial model demonstrates how most illnesses are failures of connection—with our bodies, ourselves, and others—and how one effective treatment may be to expand the social-embodied possibility space.

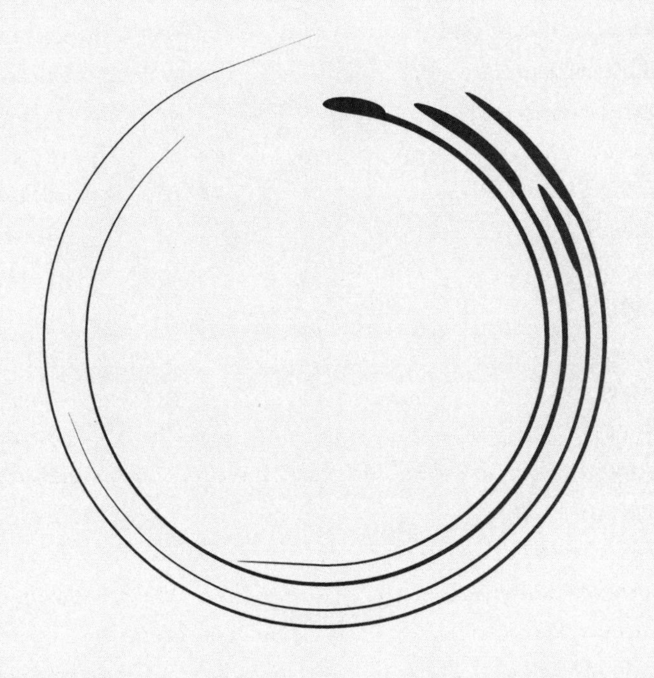

II. DISCONNECTION

Disconnection

My first few years in Berlin, I lived with two loving humans in a flat on the edge of the east part of the city, in the center of the bustle and nightlife. There were four rooms total, and new people would move in every few months, friends would visit several nights a week, and there was a real effort at building community and sharing our lives with one another. When I first moved in, it was a struggle to just be a body amongst other bodies. Without a substance, the moment-to-moment narrative of my body felt uncertain, and I became hyper-aware of my internal experience as a kind of relational time-keeper, feeling out the rhythm of an exchange. Frequently, my mind would barge in and hijack the rhythm, cutting it short because it feared my body wouldn't be able to keep it up. I'd often be on the verge of withdrawing.

In the spring of 2018, a flatmate's friend came to Berlin for a meditation retreat. As we chatted in the kitchen, I told him I'd been considering taking up meditation but wasn't sure I'd get anything out of it.

"Is anything bothering you in your life right now?" he asked.

I told him I'd had some issues with anxiety.

"I don't think meditation will help you," he said. "Anxiety is physical."

Slowly, month by month, I drank less and connected more and spent more time in my body. By June, I no longer thought of a buzz as the hum of connectedness but the hum of distance, a phone off the hook. A natural hum seemed to resonate in my bones and bring me closer to everything around me. It was the hum of what was possible, not what was certain. The more people I met, the more I wondered about this relationship—the body and the social space—and how it related to mental health. I also wondered how I'd become so disconnected in the first place.

One night in the summer, after I'd started digging into the research on psychedelics, I discovered an episode of BBC's *All in the Mind* podcast covering several topics, two of which were MDMA's potential to treat alcoholism and the use of interoception to manage autism. At first these topics might seem unrelated, but upon closer inspection I believe they may be talking about the same thing.

The MDMA story focused on the way the drug allows one's mental pathways to change, "freeing people up to think in new ways about old problems." The fact that MDMA makes us more social as a party drug wasn't even mentioned. But the whole time I was thinking, "Makes perfect sense if you see alcoholism as a disease of the social brain," for which there is considerable evidence in the neurobiological literature. Couple that with the fact that MDMA is an entactogen—a drug that stirs up feelings in the body—and you have a similar narrative to autism and interoception: allowing the body to revise its relationship with the interpersonal world.

What follows is an alternative view of what we typically consider "mental illnesses," showing how they may more usefully be thought of as social-emotional afflictions expressed through the body, with psychedelic science offering some of the support for that. I want to emphasize that, in clinical circles, diagnostic frameworks themselves are currently being brought into question, as labels like "depression," "bipolar disorder," and "ADHD" tend to mask the heterogeneous nature of these conditions. In other words, no condition arises for exactly the same reasons for each individual, and no condition manifests itself in exactly the same way for each individual. What's more, the term *disorder* itself invites some level of skepticism, as some of these conditions have recently been reframed by evolutionary psychiatrists as adaptive responses to adverse experiences, or as exaggerated, maladaptive manifestations of such responses. Still, these labels have a utility, and I'm using them here to illustrate just how inextricably intertwined the mind–body connection is with social connection.

ALEXITHYMIA

Many illnesses emerge from a disconnection to one's feelings. *Alexithymia*, a difficulty in identifying your own emotions, is implicated in just about every affliction under the sun, including substance abuse, psychosomatic disorders, anxiety, depression, eating disorders, addiction, OCD, bipolar disorder, schizophrenia, and the distress surrounding infertility. There is even one study linking alexithymia to cell phone addiction.

An emergent theory suggests that poor interoception is central to alexithymia. "Alexithymia is associated with poorer interoceptive accuracy, yet an over-reporting of subjective physical symptoms including a hypersensitivity to touch," write researchers from the University of Sussex. "These latter findings demonstrate a mismatch between objective and subjective aspects of body awareness, possibly impacting emotional processing and 'sense of self.' Indeed, alexithymic subjects show reduced emotional awareness and higher malleability of body representation in illusions of body-ownership."

The objective/subjective bit in that paragraph is the most telling: alexithymic patients report one thing, while brain scans and measures of heart rate and skin conductance paint a different picture.

One likely explanation for this mismatch is embodied predictive coding. If we become seriously disconnected from our bodily reality, we may stop listening to our bodies altogether, and instead do whatever our mind tells us will make us feel good, even if it's bad for us. In the 2008 paper "How Do Self-Assessment of Alexithymia and Sensitivity to Bodily Sensations Relate to Alcohol Consumption?" researchers wrote: "Our findings support the theoretical proposal that alexithymia is an expression of impaired processing of bodily sensations including physiological arousal, which underpin the development of maladaptive coping strategies, including alcohol use disorders. Our observations extend a growing literature emphasizing the importance of interoception and alexithymia in addiction, which can inform the development of new therapeutic strategies."

Although much more benign than alcoholism, caffeine addiction has been connected with alexithymia too. In a study of 106 male and female university students aged 18–30 years, the 18 participants who qualified as alexithymic based on their Toronto Alexithymia Scale (TAS-20) scores reported consuming nearly twice as much caffeine per day as the other students.

The common thread here seems to be uncertainty. More than anything, the brain wants to be right. If it doesn't understand how we feel, it will lead us to do things that it already knows will make us feel a certain way. If that doesn't work, it will create a predictable reality for us.

To engage in adaptive self-regulation, and prevent the various afflictions mentioned above, at-risk individuals need to be able to feel—and use—their bodily signals with some degree of accuracy. As the severity of some conditions, including suicidal ideation and eating disorders, has been correlated with the extent to which people trust and use their bodily signals to navigate daily life, alexithymia can be a serious detriment to well-being.

Alexithymia sometimes has origins in relational development: "Alexithymia is associated with interpersonal trauma during development. Parenting style, notably poor maternal care, and avoidant attachment, predict the later expression of alexithymia across patient groups."

Studies have also suggested a genetic component, as well as linkages to various types of brain damage, particularly to the insula, the region that receives interoceptive signals.

Interestingly, 60 percent of autistic individuals report having alexithymia, suggesting that interoception has a different impact on different neurotypes and may even play a role in developing one neurotype over another.

As it's not formally considered a disorder, but rather a symptom of other afflictions, there's no current treatment for alexithymia. However, in an interview with *Goop*, UCLA scientist Charles Grob, MD, said, "Individuals under the influence of MDMA have a remarkable facility to be able to articulate feeling states. So for people who are alexithymic—that is, they cannot express feelings verbally—it's thought to be a very, very valuable adjunct to psychotherapy."

SCHIZOPHRENIA

Though we traditionally think of schizophrenia as a classic "brain disease," where people are so trapped in their minds that they begin hallucinating, the more recent scientific literature paints a different picture. Schizophrenia is far more body-based, and far more relational, than you'd think.

One great mystery of psychiatry is that schizophrenia has never developed in a person with congenital blindness. "In some ways, this is one of the most interesting observations in a long time in schizophrenia research, because it's the only thing that seems to be protective against schizophrenia," Steve Silverstein, MD, a psychiatrist at the University of Rochester, told *Vice* magazine. "I think there's something here and this should be looked into much more."

There are two sides to the coin: not only is there something about the visual system that contributes to the onset of schizophrenia symptoms; there's also something about not having a visual system that reallocates cognitive resources in a way that protects against those symptoms.

In the famous rubber hand illusion (RHI), you place both hands palm down on a table with a divider between them. A rubber hand is placed next to your right hand, so that it looks like you've got both hands on the right side of the divider. An experimenter then takes a soft brush and strokes your left hand and the rubber hand at the same time, and eventually you begin to feel as though the rubber hand belongs to you. In a lab, study participants fill out a self-report questionnaire afterwards. Outside a lab, for instance at a carnival, the experiment ends in surprise, when the person with the brush strikes the rubber hand with a hammer and you jump back, tricked into feeling like it's part of your body.

People high in alexithymia are more susceptible to the illusion, struggling to integrate simultaneous sensory events into a single experience. "Higher susceptibility to the illusion in high alexithymia scorers may indicate that alexithymia is associated with an abnormal focus of one's own body," write a group of researchers at the Psychological Sciences Research Institute in Belgium.

Curiously, people with schizophrenia are also more susceptible to the illusion. But isn't schizophrenia a "mental" illness, marked by hallucinations that have nothing to do with the body? Maybe not. "The rubber hand illusion is quantitatively and qualitatively stronger in schizophrenia," writes a team from Vanderbilt University's Department of Psychology. "These findings suggest that patients have a more flexible body representation and weakened sense of self, and potentially indicate abnormalities in networks implicated in body ownership. Further, results suggest that these body ownership disturbances might be at the heart of a subset of the pathognomonic delusions of passivity."

In fact, schizophrenic patients sometimes have out-of-body experiences during the experiment, which the Vanderbilt team says "links body disownership and psychotic experiences." The effect of the illusion is explained this way: "When you hold your hand out, it's generally thought that you know it's there because of your muscles and your tendons and that sort of thing, but what the rubber hand illusion does is show how that information can be overwritten by visual information." Since the brain combines information from the senses to create a feeling of body ownership, what we see has enormous influence over what we feel and how we perceive our own body, leading to the false belief that the rubber hand is ours.

But as the Vanderbilt researchers suggest, the part of the experiment that relates to schizophrenia isn't so much the feeling of ownership over the rubber hand as it is disownership over the left hand.

Another group of researchers, from Italy, found that during the illusion, the "strength of electrical pulses that got through to [participants'] real left hand dropped dramatically. This was very surprising for us; the effect was so strong. Because the brain no longer considers the hand as part of the body, we become less able to use it." If we stop recognizing tactile input as the primary source of our bodily experience, and trust what we see instead, our brain literally disowns our body.

In the case of schizophrenia, delusions may be a direct result of this process. To further understand the point, think of what happens when you're in a darkened movie theater, engrossed in a film. In order to believe what you see, to be fully absorbed in it, you have to lose yourself. It may be the same with schizophrenia: visual hallucinations require a weakened sense of self, in particular of body ownership. The more schizophrenic people ignore actual sensory input—the feeling of the seat underneath them, the sound of people munching popcorn, the lights in the aisle—the less real those things become to the brain and the more real the screen becomes.

What seems to remedy this body condition? Social connection.

In 2005, the *Washington Post* reported on a huge but largely dismissed 30-year-study led by the World Health Organization: "People with schizophrenia typically do far better in poorer nations such as India, Nigeria, and Colombia than in Denmark, England, and the United States." Tracking about 3,300 patients in a dozen countries over three decades, the study found that patients in poorer countries "spent fewer days in hospitals, were more likely to be employed, and were more socially connected." Nearly two-thirds of the patients escaped all symptoms, compared to only about a third of patients from rich countries.

"Most people with schizophrenia in India live with their families or other social networks—in sharp contrast to the United States, where most patients are homeless, in group homes or on their own, in psychiatric facilities or in jail," writes journalist Shankar Vedantam. "Many Indian patients are given low-stress jobs by a culture that values social connectedness over productivity; patients in the United States are usually excluded from regular workplaces."

In addition, Indian families are invited to doctor appointments because "families are considered central to the problem and the solution." In America, he says, "doctor-patient conversations are confidential—and psychiatrists primarily focus on brain chemistry."

"Social connectedness for patients is seen as so important," Vedantam writes, "that the psychiatrists tell families to secretly give money to employers so that patients can be given fake jobs, work regular hours, and have the satisfaction of getting 'paid'—practices that would be unethical, even illegal, in the United States."

William Carpenter, MD, director of the Maryland Psychiatric Research Center, helped treat about 90 schizophrenia patients at three hospitals in one of the first WHO studies. He found that medications "primarily controlled patients' delusions and hallucinations, not the 'negative' symptoms that cause patients to disappear into silent, inner worlds."

Traditional psychiatry has always been about reducing psychosis, he says. But antipsychotic drugs that temper outward symptoms may actually exacerbate inward symptoms like social withdrawal: "While we treat one part of the illness, we potentially complicate another part of the illness."

The current system in wealthy countries, Vedantam writes, "merely brings patients who are in crisis into hospitals, stabilizes them with drugs, and discharges them after a few days." In the article, Benedetto Saraceno, MD, director of the Department of Mental Health and Substance Abuse at WHO's headquarters in Geneva, says this approach is "doomed to end in a new crisis—the familiar 'revolving door.'"

"Good mental health service doesn't require big technologies but *human technologies*," he says. "Sometimes, you get better human technologies in the streets of Rio than in the center of Rome."

Could it be that one therapeutic mechanism behind this relational treatment of psychosis is a gradually enhanced, bodily self-awareness? Could being seen by others remedy body disownership on some level?

STRESS

Sometimes we use our bodies to manage stress the same way we use pharmaceuticals to manage any illness: do (or take) the thing—yoga, Xanax, meditation, breathwork, Prozac—to achieve calm. Another way of looking at it is that the body isn't a vacation from the mind; the body is home, the primary reality. Being in the body feels therapeutic not because it gets us out of our stressed-out minds, but because when we're there, we realize we aren't as stressed as our mind thinks we are. We breathe, we relax, we become present. This isn't a newly achieved state of calm; we just weren't aware of it before. In some cases, poor bodily awareness may be what engenders stress in the first place.

In a study on stress resilience and interoception, neuroscientists at the University of California, San Diego found that people with low resilience to stress "show reduced attention to bodily signals but greater neural processing to aversive bodily perturbations." In other words, they are consciously experiencing how their bodies might feel in a potentially threatening future more than how they actually feel now.

The neural processing was happening in the anterior insula and the thalamus, brain areas involved in anticipating threat. The researchers interpret this as a sign of inefficient neural processing in "low-resilient" individuals. "High-resilient" people, such as elite athletes, do not show much activation in these regions in response to anticipated aversive events. It's the over-processing of threat, paired with under-processing of the body's actual state, that leads to the suffering.

They explain: "In low-resilient individuals, this mismatch between attention to and processing of interoceptive afferents may result in poor adaptation in stressful situations."

What this means is that if you experience a lot of stress, it may not be because you are actually that stressed ("you" being your body, the primary reality) but because your mind has raced ahead trying to anticipate the threat of failure, a missed deadline, or a lost opportunity. But none of that has happened yet and, underneath it all, your body knows this better than your mind.

Some level of stress is healthy, and completely necessary. In fact, the brain has evolved to protect the body, and to guide the body's response to potential stressors. That's helpful if we're judging whether or not we can safely cross the street, or if we can eat another cookie without feeling uncomfortable. But it's not helpful if we're stressing out or getting anxious about things we'd rather not stress out or get anxious about. Stress and anxiety are just the brain trying to anticipate whether or not there will be a threat (a loss, a failure, a negative reaction), and to suit up the body in the right kind of armor. They are just states of heightened alertness, meant to help us put energy into anticipated defense. But that's just it: frequently, nothing bad has happened yet, and may never happen. If you check in with how your body actually feels in the present moment, when your brain is not busy anticipating the future, then the calmness may return.

"Interoception is an important process for resilience because it links the perturbation of internal state, including stressors, to goal-directed action that can restore the homeostatic balance of the body," the authors of the study write. "Findings point to bodily awareness training as a potential intervention for those who report impaired stress resilience."

Seeking social support is one of the most effective ways to reduce stress. It can lower blood pressure and cortisol levels and give us the boost of oxytocin, serotonin, and dopamine we need to persevere and make it through our challenge. Many of us have trouble asking for help, or think we have to manage our stress alone, but we always have a choice—another possibility—before us.

LONELINESS

In 2018, in a study conducted on mice, neurobiologists at Thomas Jefferson University in Philadelphia found that too much social isolation changes the structure of the brain.

Mice raised in communities of multiple generations were divided into separate groups upon entering adulthood, with some taken out and put into cages on their own. After a month of isolation, the overall size of the mice's neurons had shrunk by about 20 percent in the hippocampus and cerebral cortex.

Interestingly, after the first month of isolation, the mice's neurons had "a higher density of dendritic spines—structures for making neural connections—on message-receiving dendrites." This finding was surprising since an increase in spines is usually a positive sign. "It's almost as though the brain is trying to save itself," said lead researcher and professor of neuroscience Richard Smeyne, PhD.

But after three months, dendritic spine density had returned to baseline levels. "It tried to recover, it can't, and we start to see these problems," Smeyne said—problems that included reduced neuronal growth and increased stress hormones.

In humans, too much isolation can lead to anxiety, depression, psychosis, and cognitive dysfunction more generally. Extreme cases include the effects of solitary confinement in prison systems around the world—and neuroscientists have identified devastating, lasting impairments in memory, spatial navigation, facial recognition, and physical health. "We see solitary confinement as nothing less than a death penalty by social deprivation," said Stephanie Cacioppo, PhD, an assistant professor of psychiatry and behavioral neuroscience at the University of Chicago, in a 2018 panel discussion at the annual Society for Neuroscience conference.

But even if we know that too much isolation isn't good for us, a healthy amount of solitude is often necessary and constructive. It would perhaps be immoral, let alone unrealistic, for neuroscientists to advise people on how much alone time is too much. In fact, researchers who study loneliness formally define it as "perceived social isolation," since it can be so subjective. So how do we know, as individuals, when we've crossed that line for ourselves?

In a meta-analysis comprising thousands of college students, Arnold and Dobkins found diminished body trust to be the key interoceptive aspect of loneliness, even when accounting for related constructs such as depression, alexithymia, gratitude, and self-compassion.

"Here we report the first known associations between loneliness and interoceptive sensibility (IS)," they wrote. "Interestingly, particularly successful attempts at loneliness reduction have used aspects of mindfulness meditation, which may train interoception, as a candidate mediating mechanism of loneliness reductions. Low body trusting, specifically, has recently been shown to predict suicidal ideation and attempts as well as eating disorder severity. Thus, this relational mechanism to core constituents of emotional information (bodily signals) may contribute to social disengagement and social mistrust in lonely individuals, maintaining the condition through misevaluation of social connection."

Learning to trust your body signals—which Dobkins calls the "physical manifestation of trusting oneself"—is just as important in sensing feelings of social hunger as it is in sensing feelings of physical hunger.

ADDICTION

Addiction is also closely tied to interoception, with addicts showing altered bodily awareness and interoceptive processing. These changes show up across the board in people with drug, gambling, food, Internet, and cell phone addictions. Martin Paulus, MD, scientific director of the Laureate Institute for Brain Research in Tulsa, Oklahoma, theorizes that interoception contributes to drug addiction "by incorporating an 'embodied' experience of drug use together with the individual's predicted versus actual internal state to modulate approach or avoidance behavior, i.e., whether to take or not to take drugs." What's more, the choice to approach or avoid a drug can be "socially" motivated, even if the addictive behavior itself doesn't occur in a social environment.

"Addiction happens when we're stressed and we need to soothe ourselves," addiction expert Gabor Maté, MD, has been known to say, along with his popular refrain, "Ask not why the addiction, but why the pain." Gül Dölen, MD, PhD, a neurobiologist at Johns Hopkins Center for Psychedelic and Consciousness Research, echoes this notion. "There is definitely evidence that substance use disorder could be considered a disease of the social brain," she told me in an interview. Some experts even put it this way: "Maybe we shouldn't call it addiction. Maybe we should call it bonding."

At the University of Zurich, Katrin Preller, PhD, studies the social-cognitive correlates of heroin and cocaine addictions as well as the social effects of psychedelics and MDMA. Expanding on the large body of research already revealing social-interaction deficits in drug addicts, her team found that cocaine users show diminished emotional engagement in social interaction via a social gaze eye-tracking task. Compared to non-addicts, they also showed reduced activity in the medial orbitofrontal cortex, a key region of reward processing, during the task. "Social interaction deficits in cocaine users as observed here may arise from altered social reward processing," her team writes. "These results point to the importance of reinstatement of social reward in the treatment of stimulant addiction."

In a series of Johns Hopkins studies targeting nicotine addiction and psilocybin, participants "identified social factors, i.e., smoking, as a way of connecting with other people, that contributed to their addiction." They reported psilocybin-induced feelings of love and connection with their environment and other people, independent of smoking as a social factor, as important for quitting smoking. "Psilocybin may have reinstated social reward processing, helping patients to overcome their addiction," Preller speculates. "My hope is that therapy will focus more on social cognition and the social environment of patients. For example, social trainings may aim at reinstating social reward processing in addicted patients, helping them to reconnect with their social environment."

Alcoholics Anonymous (AA) and other support groups are still some of the most successful treatments for individuals with substance use disorders. After evaluating 35 studies—involving the work of 145 scientists and the outcomes of 10,080 participants—Keith Humphreys, PhD, professor of psychiatry and behavioral sciences at Stanford University, and his fellow investigators determined that AA was nearly always more effective than psychotherapy in achieving abstinence. In addition, most studies showed that AA participation lowered healthcare costs.

Why do social ties matter so much for addiction treatment? Two obvious explanations might be that social support helps relieve people of the burden of dealing with their issues alone and helps hold people accountable for their own improvement, and both of these are probably correct. But these are also mind-based ways of looking at it. Is there more going on than this, on a physical, body-based level? When we talk about "social support," what we're really talking about is a body-based feeling: a feeling of safety and homeostasis, that there are people around to catch you if you fall, along with the reassurance—at least in the case of AA—that one can easily find a meeting, even when traveling to other cities or countries.

The Bristol Imperial MDMA for Alcoholism (BIMA) study has also shown that MDMA is effective in treating alcoholism. As an entactogen that makes people feel more social, could there be a link between the social feelings of MDMA and the social ties of AA? Maybe it's the case that MDMA isn't doing anything new, but rather showing us why AA works so well. Typically, the effects are explained in terms of enhancing the patient–psychotherapist bond, but it also appears that there's a window of social learning that is reopened neurobiologically on MDMA. Could the same thing be happening at AA meetings, over time?

"Social support seems to affect our balance of hormones. Adequate amounts of social support are associated with increases in levels of a hormone called oxytocin, which functions to decrease anxiety levels and stimulate the parasympathetic nervous system calming down responses. Oxytocin also stimulates our desire to seek out social contact and increases our sense of attachment to people who are important to us."

When you consider that many of us use alcohol to calm the nerves, or to manage feelings of grief or separateness, it might be plausible to suppose that feeling socially safe may serve as a replacement for the drug. MDMA acutely enhances that feeling of safety and puts us in a more open state, so that we are more likely to allow ourselves to feel things we'd otherwise suppress. But perhaps the right social support can elicit the same effect.

"I definitely think that the [addictive] drug gains in power and reward value," says Harriet de Wit, PhD, a behavioral pharmacologist at the University of Chicago, who speaks to me in a later part of the book. "So then by comparison, other rewards are less potent. And if you can kind of increase the value of other rewards, then you might be able to override the drug rewards. There's recently been a lot of interest in drug reward versus social reward. There's a group in Baltimore that has been giving animals the choice of a drug reward and a social reward, and the social reward wins hands down."

ANXIETY

In a Medium post about alcohol and stress, Michael Sayette, PhD, an alcohol researcher and professor of psychology and psychiatry at the University of Pittsburgh, is quoted as saying, "If you're consuming quite a few drinks to reduce stress and you do this all the time, what happens is that you start to alter your brain chemistry. The brain's normal non-alcohol state begins to change, and in some cases it may become a more anxious one."

This new anxiety baseline can fuel further drinking, says the author of the post, and so there's a snowball effect where stress and alcohol feed off each other. Alcohol might help immediately relieve stress, explains Johns Hopkins neuropsychopharmacologist Manoj Doss, PhD, but "your subsequent sober state can later become a more anxious one."

In terms of interoception, anxiety-prone individuals may not have a heightened baseline interoceptive state, but rather an exaggerated expectation (i.e., belief) about future body states.

Describing the "somatic error" hypothesis of anxiety, wherein a person's predicted body state is out of line with their current body state, researchers at Kent State University write:

> In the context of this framework, anxiety sensitivity can be conceptualized as manifesting from two primary disturbances: (a) hyper-precise priors that generate exaggerated body predictions whenever exposed to sensory signals that have become associated with anxiety, and (b) the persistent triggering of abnormally large somatic errors by sensory signals (either real or imagined) irrespective of context—that is, context rigidity. These sensitized sensory triggers lead to somatic error, followed by corrective action that upregulates the observed body state to be more in line with the predicted body state. However, by upregulating the very visceral signals one is sensitized to, the process can quickly escalate into a vicious cycle that culminates in the avoidance of all sensory cues associated with the experience of anxiety.

One unfortunate consequence of anxiety disorders, then, is another kind of withdrawal: avoidant behavior.

> Although avoidance of perceived triggers quickly reduces the somatic error, the relief is only temporary, and the vicious cycle returns whenever the trigger is once again encountered. Because the ultimate goal of the corrective action is to minimize somatic error as quickly and efficiently as possible, whether or not the corrective action is adaptive or maladaptive is not considered during the regulatory process. For this reason, avoidance and withdrawal behavior is often the action of choice because it rapidly reduces somatic error. As the error is reduced, the avoidance behavior is reinforced, which helps to explain why avoidance is the primary behavioral manifestation of anxiety and the symptom that elicits the most impairment in life functioning.

At the Max Planck Institute of Neurobiology in Munich, Nadine Gogolla, PhD, is one of the world's leading experts on the insular cortex. Her work has encouraged a shift to an "insular view of anxiety" that further confirms the close ties between the body and social emotion:

> There is some evidence, from functional imaging studies of the

> brain, that the insular cortex is hyperactive in people suffering from anxiety disorders compared to healthy controls and to other structures in the brain. We see this right across the board in people with PTSD, those who suffer panic attacks, generalized anxiety disorders, phobias, and social anxiety. In all these conditions, the activity of the insular cortex has been found to be altered. Metabolic studies and post-mortem studies have also found differences in the insular cortex, for instance, in the expression of neurotransmitters released by neurons in this part of the brain.

The insular cortex is one of the brain regions that most consistently operates differently in people with anxiety disorders. "I strongly believe that understanding the basic physiological role and function of the insular cortex," Gogolla says, "can give us insights into what goes wrong in anxiety disorders that can open doors to finding new ways of tackling them."

At University College London, Sarah Garfinkel's team is currently doing research for a training using interoception as treatment for anxiety. "Anxiety is associated with interoceptive error," Garfinkel explains. People with anxiety think they're good at detecting bodily changes, and focus a lot on themselves. "Training people to become more accurate, especially at their heartbeat, does produce drops in anxiety." For example, it may help if you have "early access" into a signal like your racing heart and can notice it and calm yourself. Otherwise, it might be too late and you're already in an anxious state before you notice. The goal is to notice bodily signals, regulate your reaction, and not worry about them.

You need three things to reduce anxiety, Garfinkel says: 1) good accuracy to reduce error and help facilitate spontaneous regulation; 2) focus on the outside world and not having interoceptive bias; and 3) not worrying about those body signals when you detect them.

Not worrying, of course, is easier said than done, but perhaps this is where Dobkins and Arnold's "beginner's mindset" could come in handy.

DEPRESSION

There are as many causes for depression as there are individuals with depression, but there are some commonalities in the way it arises. Traditionally, depression has been considered a "mental" illness, characterized by rumination and destructive thinking patterns. Another view is that depression has far more to do with the body than we think, and the clinical research is beginning to substantiate this perspective.

There are more and more body-based, somatic treatments being integrated into depression treatment programs. We know that exercise helps, for example. And massage. But when we try to unearth explanations for why these physical treatments work, we are still focused on neurochemistry as though it's all about releasing the right cocktail of chemicals in the brain.

The *New York Times* reported in 2020 that even "light activity—walking at a casual pace, shopping, playing an instrument, doing chores around the house—has a big effect [on mental health]." The explanation was as follows: "biological mechanisms, reducing inflammation, promoting self-esteem, social support, self-efficacy."

When we dig a little deeper into those biological mechanisms, we are still focused on the mind: exercise releases endorphins, dopamine, endocannabinoids, serotonin; then we feel happy. None of the aforementioned explanations considers our *relationship* to our body, specifically, as being important.

"Movement is intimately connected to the brain's creation of a sense of body ownership," write Luke E. Miller, PhD, and Alessandro Farnè, PhD, at the Lyon Neuroscience Research Centre. "Movement, it seems, is the 'glue that binds the body with the self.'"

Researchers at the Milan Center for Neuroscience agree. "Our novel findings reveal a dynamic and plastic contribution of the motor system to the emergence of a coherent bodily self, suggesting that the development of the sense of body ownership is shaped by motor experience, rather than being purely sensory." At the same time, "impairment of the motor system directly affects the multisensory sense of body ownership."

What if it's the sense of body ownership itself—the feeling that you own the body that's powering through a marathon training run, holding a yoga pose, dancing at a concert—that releases those chemicals? What if the brain is designed to register connection to the body as a reward, if we allow ourselves to connect? Know your body, know yourself. There is no way to be uncertain of the fact that you are moving. In fact, some neuroscientists recommend taking a walk to manage a bad psychedelic trip, specifically because psychedelics do not affect the motor cortex as much as other parts of the brain. The body grounds us.

Researchers at the University of Zurich found that depression has a lot to do with the brain's capacity to successfully regulate bodily states. Depression doesn't come from your external circumstances, or your thinking style, so much as your brain's observation that you aren't doing anything to change your situation based on the signals your body is providing. It's about listening to those signals and doing something with them.

Let's not forget that breathing is a motor function too. In a study on lovingkindness, focused attention, open monitoring, and mantra recitation meditation practices, Millière and Carhart-Harris write that the involvement of the insula in all four types of meditation points to "the central role of the attentional control of bodily awareness, and awareness of breathing in particular, during various contemplative practices," shedding some light on their therapeutic effects.

At the same time, I suspect that another simple way to restore our sense of self and body ownership is for other people to recognize us. That may be one reason why love and intimacy feel so good: we are confirmed by another. Our body is acknowledged as real. Social touch and group connection could also be powerful in treating illnesses like depression, schizophrenia, addiction, and anxiety for this reason.

This is already suggested in the neuroscience literature: "Body-ownership and the sense of agency interact at the level of the left middle insula, a high-level multisensory hub engaged in body and action awareness in general." In other words, it's one spoke in the "social interoception" hub of the brain.

PTSD

The anterior right insula seems, in particular, to be where our felt sense is produced. Michael Jawer in *Psychology Today* writes:

> The right anterior insula is also the part that corresponds most closely to the severity of a person's PTSD symptoms. When someone is having a fearful flashback, that part of the insula is highly activated; when someone is feeling distant or nothing at all, that part of the insula shows very low activation. A

> person's felt state, therefore—one's very sense of self—is tangibly diminished when he or she is in the throes of dissociation. The flip side is that people who are "tuned in" to their bodies (and who, consequently, are more emotionally attuned) actually have more developed right anterior insulas, as measured by the amount of gray matter residing there.

Some therapy techniques, such as interoceptive exposure, target this part of the brain. Originally designed to treat panic disorder, interoceptive exposure is designed to help people directly confront feared bodily symptoms often associated with anxiety, such as an increased heart rate and shortness of breath, by gradually re-exposing them to those feelings. The therapist may facilitate this by having a person (in a controlled and safe manner) hyperventilate for a brief period of time, exercise, breathe through a straw, or hold their breath. There is evidence that interoceptive exposure may be successful in treating PTSD as well.

In a 2020 Cornell University study, researchers found that "interoceptive brain capacity enhanced by Mindfulness Based Stress Reduction appears to be the primary cerebral mechanism that regulates emotional disturbances and improves anxiety symptoms of PTSD."

If we maintain the view that feeling states are dynamically impacted by our environment, however, PTSD becomes a relational issue. At the moment, PTSD is the main clinical target of MDMA trials. How might MDMA work to treat PTSD? Could it share a common treatment mechanism with other applications, such as MDMA-for-alcoholism? Most trauma is interpersonal. It therefore makes sense to use a social drug to heal it. If you read the media literature on MDMA-assisted therapy, you'll find that most researchers claim the drug itself does not work to "cure" PTSD, but rather it helps facilitate the therapy process by improving the bond between the patient and therapist. Be that as it may, there are many documented and anecdotal cases of individuals healing from various conditions on their own using these compounds. We can't ignore the validity of these accounts. They are personal science. But are they "interpersonal" in nature, regardless of the solo setting?

A study on 765 Vietnam War veterans in New Zealand found that PTSD adversely affects veterans' interpersonal relationships: "It is suggested that higher levels of PTSD affect the ability of veterans to initiate and maintain interpersonal relationships and that these interpersonal problems are evident in poorer levels of family functioning and poorer dyadic adjustment."

While the causes of veteran PTSD—meaning, the traumatic event(s) that occurred—may not necessarily be interpersonal themselves, their effect on a veteran's nervous system is inseparable from their bodily, emotional, and therefore social experience of the world.

Results from clinical trials using MDMA to treat PTSD in vets suggest that this healing is interpersonal in nature.

Although veteran PTSD is just beginning to be framed as interpersonal in this way, complex PTSD (c-PTSD), another common disorder resulting from chronic exposure to traumatic circumstances rather than a single event, is typically classified as interpersonal and frequently tied to childhood experiences. Interestingly, some researchers have found that c-PTSD responds a little differently to psychedelic-assisted treatment than PTSD.

"The treatment of c-PTSD required larger numbers of psychedelic experiences than the MDMA-assisted treatment of a single trauma that is currently being investigated in phase 3 studies," wrote Swiss clinicians Peter Gasser, MD, and Peter Oehen, MD, in a 2022 paper examining the effects of MDMA and LSD group therapy on c-PTSD. MDMA sessions ranged from 1 to 9 applications, and 1 to 12 for LSD. "Considering that such disorders often begin early in life and have a significant impact on personality development, this is not surprising."

Again, we have an example of a condition in which psychedelic-assisted therapy may be targeting more than one's cognitive sense of self, reaching deep into the roots of the bodily self.

EATING DISORDERS

In a 2020 paper on interoception and anorexia, researchers wrote:

> The extreme eating restrictions characterizing anorexia nervosa may serve the fundamental role of keeping under control (i.e., reducing excessively high levels of) interoceptive uncertainty, above and beyond controlling and mitigating concerns for body weight. In other words, noisy interoceptive streams may instantiate active strategies (i.e., eating restrictions) that amplify autonomic hunger signals to minimize interoceptive

> uncertainty and maintain a more coherent sense of (interoceptive) self.

In other words, the instigating problem may not be that folks with anorexia have body image issues; it may be that they're uncertain how it will make them feel to eat, and feeling hungry is predictable and safe.

The same theory applies to overeating. In another 2020 paper, on emotional eating in adult obesity, researchers drew a related conclusion:

> The results suggest that good interoceptive awareness can increase the risk of emotional eating if not supported by good interoceptive reliance. Interoceptive reliance, like the ability to trust, positively consider, and positively use inner sensations, should be a privileged target of psychotherapeutic interventions in obesity.

This "trust" is key, and hints at a facet of interoception that frequently predicts the onset of eating disorders: diminished body trust. As a dimension of interoceptive sensibility, body trust is defined as the degree to which a person feels safe and at home in their body, as well as how much they trust their bodily signals. Studies have identified diminished body trust (or interoceptive reliance, as some researchers phrase it), and particularly "not feeling safe in one's body," as the central bridge between interoception and eating disorder symptoms.

Another study, looking at anorexia (AN) versus bulimia (BN), suggests involvement of the insula and interoception in both disorders.

> Although AN and BN share several overlapping symptoms and traits (e.g., anxiety, body image disturbance), they are also characterized by divergent traits (e.g., avoidance versus approach; inhibited versus impulsive) that might suggest opposite patterns of response to the anticipation and experience of interoceptive stimuli.
>
> A "healthy" psychological experience of our own body is central to mental health, as revealed by eating and psychosomatic disorders, but also the somatic dimensions of other conditions such as depression and anxiety. Therefore, rebalancing the relationship between internal and external bodily signals

> can help not only develop treatments specific for AN patients, but also to prevent the development of such disorders in young people, targeting ages when they might be more sensitive to bodily as well as emotional changes (since body trust has been shown to dip temporarily during adolescence).
>
> These findings are particularly important not only because the sense of body ownership is a fundamental aspect of our bodily self-consciousness but also because the representation of one's own body is not fixed but rather the result of a predictive processing that is constantly updated by means of multi-sensory processes.

What's more, the right anterior insula has consistently been implicated in anorexia nervosa. This part of the insula is thought to play a very specific role in social cognition: helping us understand other people's feelings and bodily states rather than their actions, intentions, or abstract beliefs.

Comparing groups of women with and without anorexia, a research team led by Timo Brockmeyer, PhD, at the Center for Psychosocial Medicine in Heidelberg, Germany, found that AN is associated with difficulties inferring others' emotional states "despite largely intact nonemotional mental state inference." Feeling into another person's emotional experience is not the same process as reading their mind, and participants in this study struggled with the former.

By some accounts, anorexia is a denial of one of the most basic human needs: desire—whether for sex, love, or, in this case, food. The logic surrounding it often can't be explained to those who don't suffer from it, and, as mental health journalist Marianne Apostolides writes, "efforts to provide rational reasons for its development seem incapable of changing the behavior."

Apostolides, who suffered from anorexia herself, thinks this may be one of the reasons psychedelics can help treat the condition. "We can talk about rumination, cognitive inflexibility, inability to process emotions," she says. "But what is felt, in the body, is the brutality of denying the body's fundamental appetites." During a psilocybin or MDMA session, a person "doesn't feel trapped in the cage of logic; she can slip into a different kind of understanding, where the dynamics underlying the disorder are made visceral and vivid. In my experience, those insights make more sense than any rational, word-based explanation."

SUICIDAL IDEATION

The link between suicide and interoception is strong as well. In one study published in *BMC Psychiatry*, researchers found that it's the tendency to ignore (not act on) bodily signals that exacerbates this condition.

> Suicide ideators do not lack the ability to perceive their own bodily signals, but they do not use them properly. They report less interoceptive sensibility, suggesting that they use this information less, in terms of a reduced ability to regulate body-related attention or use body sensations for distress regulation.

In other words, suicide ideators show diminished body trust, not unlike people with eating disorders. Suicidal ideation also has close ties to regions of the brain associated with social cognition.

In March 2020, the Allen Institute in Seattle became the first research institution to record electrical activity from the von Economo neurons (VENs) in live human tissue. An important part of the insular cortex, von Economo neurons are involved in empathy and have dramatically increased over the course of human evolution. They are found mostly in highly social species and thought to be crucial for prosocial behaviors. Also called spindle neurons, these long, cigar-shaped cells exist in only two parts of the brain: the anterior cingulate cortex (ACC) and the frontal insula (FI), which together make up the salience network (SN) of the brain. Like mirror neurons, they seem to play a central role in self-awareness, "enabling humans to see themselves as others see them, which may be an essential ability for self-awareness and introspection."

Evolutionary neuroscientist John Allman, PhD, says that it was by studying the von Economo neurons that he realized "self-awareness and social awareness are part of the same functioning." Allman thinks these neurons expedite communication from the ACC and FI to other parts of the brain, like the central executive network and the default mode network. The cells have branches that extend far across the brain and are unusually large—and in the nervous system, size often indicates speed. "They're big neurons, which I think do a very fast read of something and then relay that information elsewhere quickly."

Despite their dynamic function, they appear to be quite fragile. For example, alcoholics have 60 percent fewer von Economo neurons than average, spurred by a loss of insula gray matter.

In the healthiest humans—called "superagers"—VENs are more plentiful, helping these individuals preserve their memory capacity and stave off cognitive decline.

But there's a paradox at play. In 2011, the neurons were labeled "suicide cells" by *Scientific American*, as people who take their own lives tend to have a greater number of them, more densely packed.

In the *Scientific American* study, all suicidal subjects had psychiatric illnesses like schizophrenia or bipolar disorder. If we view these not as mental health issues but as relational health issues, it's possible to guess our way through the maze: could it be that the more von Economo neurons people with these illnesses have, perhaps as an overcompensatory mechanism, the more social connection they need? Unfortunately, it's those who need it most (schizophrenics, bipolar folks, etc.) who often end up receiving it the least, which may lead these people to entertain suicidal thoughts. The Allen Institute's work is just the beginning of a long line of research exploring the significance of these mysterious, yet clearly important, specialized neurons.

Perhaps the best-known psychedelic-type drug being used to treat suicidal ideation is ketamine. Technically an anesthetic dissociative, causing what some might call mind–body "disconnection," it can relieve suicidal ideation and depression at medium to high doses, with long-lasting therapeutic benefits. Dissociative effects are not at odds with mind–body connection, however, and what's interesting about ketamine is that awareness of the experience seems to undergird the therapeutic effect.

"I always tell people who come to treatment that it is not a linear relationship between dosage and effect with ketamine," says Sergio Perez Rosal, MD, medical director of OVID Health Systems, a ketamine-augmented psychotherapy clinic in Berlin. "With ketamine, the relation between dosage and effect could be described in a graph as a kind of inverted-U, where you start at zero, with no ketamine and no effect, then you go up and you have more ketamine and more of an antidepressant effect. At some point, it drops down and more ketamine becomes less effective. Why? Because it becomes more anesthetic. The patient starts losing their memory and the potential to remember what happened in the session. Remembering is a core part of the psychotherapeutic process."

This might be the biggest difference, he says, from serotonergic substances, which seem to have a more linear relation between dosage and psychedelic effects. "With ketamine, you need to have some level of awareness and be able to remember what happened in-session for the treatment to be efficacious long term."

What bearing might interoceptive awareness have on the dissociative effects of ketamine, and what bearing might these effects have on interoceptive awareness? Do ketamine patients go on to trust and use their bodily signals more often, perhaps expanding their relational existence more generally? If so, could this help explain ketamine's therapeutic value? So far, no one has investigated these questions.

> Over the course of 2019, I started remembering the ways in which my bodily self had developed as a kid, and I seemed to be reliving the traumas surrounding it. It was as though I'd been blind to it for 20 years, looking but not seeing, until I decided to peer through a bodily lens. It's no coincidence that my body woke up when I was halfway across the world, in another life, with another family. Gradually, I felt an intense need to spend some time with the person I felt safest with in the world, someone I'd met back in my drinking days, who would prove to be a better friend than I could have possibly imagined—my home away from home, embodied in another person. It would be during this visit to Seattle, where I'd once predicted my way into oblivion, that I had the insight to stop drinking completely. When I returned to Berlin, questions began rolling around in my brain. If disconnection from one's body, self, and social contacts leads to illness, how can we use connection to promote wellness? What does it mean to be connected, in body, mind, and spirit? Over the course of 2020, as disconnection became a familiar theme, I conversed with researchers around the world to find some answers.

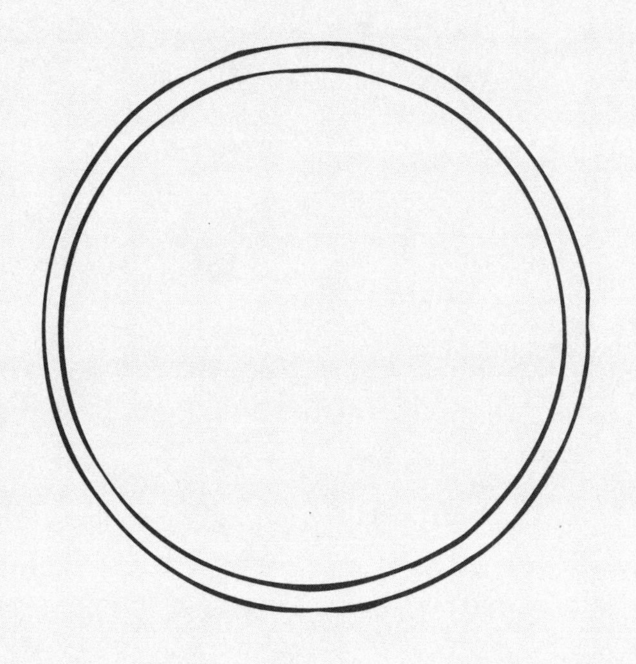

III. CONNECTION

Connection

In late February 2020, I was at a club in Berlin, wandering through a room full of people lounging on couch cushions around a luminous indoor pool, feeling synergy after a particularly lonely month. I'd taken a small bit of MDMA and could feel the essence of each of my friends. Toward most others I felt calm and safe, the way I'd felt when, earlier that week, a dear friend had visited me on her way from India back to New York and we'd huddled under an umbrella together, making our way through the rain to find some food when she arrived. This time the bodily feeling came with an insight: I had the sense that it was safe to approach and be approached by others. There was no need for defenses, specifically because I trusted myself so much that I felt I could handle anything. I had the thought, "Compassion is the only defense I need." Contrast this with alcohol, which simply dissolved my boundaries. On MDMA I didn't feel like I dropped my boundaries; I felt like I was more in touch with them and could therefore manage any kind of approach. We often hear about being connected to some higher power during the psychedelic-type experience, but the feeling of having faith in yourself, and suddenly realizing you can navigate the world from this wellspring of safety within, is a deeply spiritual experience. The person I was happiest to run into that night was C, a friend I'd been seeing for a few months whose spirit vibrated on the same youthful frequency as mine—just a couple of kids trapped inside adult bodies. When we spotted each other, my body felt pure, made of light, and I threw my arms around him. A few weeks later, just after C helped me move into a new flat in West Berlin, the first lockdown began.

AUTHENTICITY

"At the simplest level, we suggest that low quality social connection, or 'disconnection,' arises from lack of authenticity within the self."
—Karen Dobkins, PhD, neuroscientist, University of California, San Diego

One thing I appreciate about Berlin is its authenticity. The transparency and directness of German culture has something to do with it, but the city itself transcends its own national culture to a degree I've rarely seen elsewhere. Whenever I've tried to inflate or sell myself, as Americans sometimes do, I've been met with vaguely offended confusion. The entire aesthetic of the city itself is assertively unadorned, defined by walls of graffiti and repurposed factories and warehouses. It's not that this particular attitude-aesthetic is more authentic than one with more affect or ornamentation—just that it deliberately represents a nonattachment to, or non-privileging of, any particular aesthetic, which helps sustain a culture of tolerance, diversity, and inclusion. Maybe this aesthetic has itself become an image in the past 30 years or so (being turned away from a club by "dressing to impress," for example), but many of the people I've met here have been courageous in a way that has more to do with their own values than anyone else's. "Do you want to go to the park?" a new friend said one Friday night, in the middle of after-work drinks at our coworking space. "I don't feel like getting drunk." This was 2018, before "sober-curious" was an acceptable concept, and well before I myself had the courage to say such a thing.

Still, many of us—myself included—make unhealthy sacrifices for connection from time to time. Or connect inauthentically with people, like the classic scenario of having many social contacts but few true friends, leaving us feeling empty when it seems we ought to be fulfilled.

Dobkins and Arnold, who study interoception in a social context, put it this way:

> At the simplest level, we suggest that low quality social connection, or "disconnection," arises from lack of authenticity within the self. If one is not connected with what is actually happening inside oneself during a social interaction, how can there be true connection/interaction with the other person? Moreover, if one is not aware of the lack of true connection during a social interaction, one will continue to be socially hungry. Like the analogy to physical hunger, being unable to accurately read one's social hunger pains will lead to social malnutrition/starvation.

Recognizing the complexity of authentic connection, addiction and trauma expert Gabor Maté, MD, often speaks about the conflict between our need for attachment and our need for authenticity. "Authenticity means we know who we are, we know what we feel, and we are able to express it and honor it in our behaviors." But, he asks, "What happens when, in order to attach, we have to suppress our authenticity?"

The key is to honor both by learning to assert our boundaries.

"I worry about people who are nice all the time," he says. "Healthy anger is nothing but a defense of your boundaries. If you don't know how to say 'No' when you need to, your body will do it for you in the form of illness. To prevent illness, you need to learn to say 'No' and assert who you are."

In her essay collection *On Lies, Secrecy, and Silence*, the poet Adrienne Rich wrote that, in relationships, "We often use lying as a hedge against the discomfort of truly being seen." By "lying" she means suppressing the truth about how we really feel, and therefore who we really are. The primary motivation for this suppression is to avoid the immediate discomfort of being challenged or judged, but the trade-off is a chronic internal dissonance. Being inauthentic not only alienates us from others but also "engenders the greatest loneliness" by distancing us from ourselves. "Lies are usually attempts to make everything simpler—for the liar—than it really is, or ought to be," Rich wrote. "In lying to others we end up lying to ourselves. We deny the importance of an event, or a person, and thus deprive ourselves of a part of our lives. Or we use one piece of the past or present to screen out another. Thus we lose faith even within our own lives."

One way to explore the possibilities between people is to cut the bullshit, and stand firmly in whatever sludgy deluge of discomfort may follow. The alternative—fulfilling other people's expectations—feeds the prediction machine but eats away at our souls slowly over time. Which reality would you rather inhabit?

When you tell the truth, you not only honor yourself but you also create the possibility for more truth around you. Truth may drive some people away, but it will certainly bring some people—perhaps the right people—closer. All of this to say, connection must be authentic to improve our well-being with any significance.

> As the pandemic reached Berlin, and life as we knew it took a deep, uncertain breath, the only thing that seemed to matter anymore was

real connection, and being real about connection. People became kinder to each other, said the things they'd been making excuses not to say, reached out to old friends and lovers and family members, and helped those in need with acts of generosity. Would we wake up when it was over, I wondered, and think, "What an epic drunk text?" Or would we continue to be this gentle, this vulnerable, this human? Authentic connection comes in many forms, and feels different for everyone, but the following practices, theories, and phenomena seem to share a degree of universal value, and underscore the body's role in experiencing and bringing it about.

KAMA MUTA

In the middle of a writing spree in March, I googled "What does human connection feel like?" Not because I'd forgotten (although lockdown did seem to last an eternity), but because I started thinking about the fact that it's actually hard to describe. We can cite many things we do that make us feel connected to ourselves and others, but we don't spend much time trying to put words to the feeling of connectedness itself.

After eight pages of barren search results, I came across a research article by UCLA psychological anthropologist Alan Page Fiske, PhD, who studies an emotion called *kama muta*, defined as the feeling of being moved by love. As it turns out, Fiske wrote an entire book in 2019 all about kama muta, "the connecting emotion."

Kama muta (Sanskrit for "being moved by love") is a brand-new term for an ancient social emotion, one that most of us have experienced but struggle to describe. In his book *Kama Muta: Discovering the Connecting Emotion* (2019), Fiske gives language to the intense, sudden feeling of connectedness that can warm the heart, give us goosebumps, or move us to tears. His research includes surveys with over 10,000 participants in 19 nations and 15 languages. What his team has found is that kama muta is a universal human emotion, and one that we may be able to cultivate for our well-being even in the absence of social contact.

Although there is no exact word in everyday language for this emotion, Fiske says, English speakers might call it *being moved, touched, connected, awestruck, warm and fuzzy, one with humanity or the universe.*

"However, none of these terms captures precisely what the emotion is—and using any one of them conceals the fact that though it has many names, it is one emotion. So we coined a scientific term for it, 'kama muta,' borrowed from the ancient Sanskrit where it meant 'moved by love,' written in the beautiful Devanāgarī script as काममूत."

Kama muta is usually accompanied by some combination of the following: a warm feeling in the center of the chest; moist eyes, tears, or weeping; a "choked-up" feeling of a constricted throat or "lump in the throat"; difficulty speaking, or speaking in a creaky voice; chills, thrills, or goosebumps; a deep breath or pause in breathing; a phatic exclamation—an emotive marker—such as *awww*; moving one or both hands over the center of the chest; and, if it's an especially strong experience, may be followed by a feeling of lightness, buoyancy, or exhilaration.

Fiske and his team have only been studying kama muta for a few years, so the psychophysiology of it is still unknown, but a description of warmth in the heart, in particular, has characterized the sensation for millennia.

To investigate where this warmth might be coming from in the body—inside the chest, for example, or on the skin—Fiske conducted pilot research with several participants, showing them videos evoking kama muta and asking them to note when they felt the warmth. Using a sensitive thermal video camera that recorded skin temperatures at a resolution of approximately 0.1°C, he kept his eyes trained on the video. He found no detectable change in skin temperature anywhere on the chest in moments when participants reported this sensation.

"So where is this warmth?" he writes. "And if something truly is warming up in the chest, how would we sense it?" If the sensory system is simply "spoofing" warmth in the chest, or creating a kind of "referred pleasure" the way we sometimes experience referred pain, what neural systems might be responsible? This question remains a mystery, he says. "If researchers can solve this mystery, perhaps we could devise sensors that would detect this sensation and provide a biomarker of the kama muta emotion. If this becomes possible, and if this biomarker occurs only in kama muta, it would be the first physiological signature ever discovered for any emotion."

What is known, however, is that most humans experience kama muta, regardless of culture, background, personality, or spiritual orientation. Hundreds of historical sources and ethnographies support this finding, as well as observational research in churches and mosques, in poetry lounges and memorial sites, at Alcoholics Anonymous and eating disorder residential treatment programs, in birth centers, and with new parents.

Knowing how influential it can be, many humans even cultivate it deliberately, either for themselves or on behalf of others.

"Many social practices have culturally evolved via their capacity to evoke this appealing emotion," Fiske says. "The more a form of worship, a type of music, or a narrative evokes kama muta, the more people seek it out, tell others about it, and reproduce it. When a Pixar movie, a wedding practice, or poetry or photographs evoke kama muta, they spread across the globe. Preachers, orators, marketing creatives, and political consultants who can create pitches that effectively evoke kama muta are more successful than those who cannot."

When people experience kama muta, their lives improve. Not only can it strengthen relationships with friends and family, but Fiske suggests it may even help people recover from conditions like eating disorders and addiction. What's more, when communities experience kama muta, society improves. Fiske says immigrants who have kama muta experiences with people in their host country may feel a stronger sense of belonging, and natives of the host country may become more likely to embrace them.

One statement Fiske makes is especially poignant considering the circumstances at the time I discovered his work: "When people are isolated and vulnerable, excluded and distressed, kama muta can reconnect them."

In April, I emailed back and forth with Fiske to learn a bit more and tease apart the relationship between kama muta, connectedness, and peak experiences on psychedelics.

SB: What is the difference between kama muta and connectedness?

APF: "Connectedness" is not a scientific construct, so it means whatever anyone uses it to mean—which may not always be the same thing. So I can't say how it's related to kama muta, except that in many dialects, people use *feeling connected* to label communal sharing (CS) relationships. People notice the CS feeling when it's suddenly especially salient, that is, when they feel kama muta.

SB: Where does kama muta occur in the body?

APF: There are several sensations that people often have with kama muta. The most distinctive is a warm or "fuzzy" feeling on the left side of the chest; it's probably the feeling that leads people to describe an experience as "heartwarming." People often have moist eyes or tears, and often goosebumps or "chills." You may get a lump in the throat, so you're "choked up." Extraordinarily strong experiences of kama muta make people feel light, as if they were floating in the air. People vary in the sensations that they have with kama muta and may have different sensations on different occasions. Also, people apparently differ in how prone they are to physical sensations of kama muta and how attentive they are to them, and certainly not everyone is equally ready to acknowledge these sensations. A person may experience kama muta without any physical sensations, especially if the kama muta is mild. There is also an "automatic" gesture that may occur with kama muta: people often put the palm of their hand on their chest. In many cultures, this has become a ritualized gesture indicating friendly intentions, sincerity, or patriotism. Tears, chills, and a lump in the throat can also occur with other emotions. But you know it's kama muta that you're feeling if the emotion is evoked by a sudden sense of increased closeness and connection, and if it makes you want to hug and care for and express your love to people.

SB: As I was reading your book, I wondered if the feeling of "connectedness" many people report during psychedelic experiences could be another example of kama muta, especially since it usually comes as a sudden insight/feeling. If kama muta were correlated with this type of empathy/connectedness, your team might be interested in the neural correlates of the psychedelic experience.

APF: I agree that some psychoactive drugs, especially MDMA, either evoke on their own or facilitate kama muta. I discuss that on pages 244–5 and in online note 11.1 of my book. Whatever the neurochemistry of kama muta is—OT, AVP, mu-opioids, serotonin—some psychedelics, ingested under some circumstances, trigger it, in whole or in part.

SB: How can people use kama muta to feel connected during a pandemic?

APF: The pandemic has certainly generated extraordinary acts of kindness, surprising new sorts of solidarity, and signs of compassion. Kama muta is surely evoked by neighbors checking on each other, old friends reaching out over the Internet, strangers shopping for the elderly, musicians playing together from their balconies, and the evening events of clapping to applaud healthcare workers from windows and balconies. I imagine, though I don't know, that artists of all sorts are creating works that reflect these and other kama muta moments. I don't know that I have concrete tips, other than the obvious: be kind, be compassionate. We all need to connect in communal sharing moments.

> Our exchange got me thinking about my recent time in Seattle. My friend and I had been out drinking beers and playing arcade games in Capitol Hill, and there was a moment toward the end of the night when the bartender offered us shots. We considered briefly, and I saw the moment with such clarity, almost a bird's-eye view of the choice hanging in the air. I decided that I was done. There was no reward value in it anymore. In fact, I was more interested in the things we might be doing if we weren't out drinking, even if that just meant staying at home and doing nothing. The nothing—that space—suddenly seemed full of more possibility than what we were doing then, or what we could be doing later, while drinking. My friend supported my choice, and even joined me in my sobriety while I stayed with him.
>
> But what allowed me to quit for good—not just for a few weeks, as I'd done before—was the quality of our connection. Every night when I crashed on his couch and we said good night, I felt this warm, beaming sensation emanating from my heart and chest and core. It was not the heady rush of lust, nor the steady depth of partnership, nor the high of seeing a favorite friend. This was different. It came, and it went. But the connection was unmistakably in my body—a deep humility and gratitude and affection—and I felt it even more powerfully there than in my mind. The more I paid attention to it, the less I needed its replacement.

PEAK AND TRANSCENDENT EXPERIENCES

William James first described them as "mystical experiences" nearly a century ago in his book *The Varieties of Religious Experience.* In the 1920s, determined to secularize their definition, Abraham Maslow rechristened them "peak experiences," since his work found that many people from all walks of life and with very different belief systems experienced them. Now, researchers in a variety of fields—social psychology, anthropology, neuroscience—study these experiences, using a range of terms including "transcendent experiences," "self-transcendent experiences," and "kama muta moments." At their core, they are sudden moments—or extended moods—of connectedness to a higher power, the world, universe, other people, and/or ourselves.

In his book *Transcend: The New Science of Self-Actualization*, humanistic psychologist Scott Barry Kaufman, PhD, uses the terms *peak* and *transcendent* experience interchangeably to describe a feeling of unity with everything, which might include other humans, nature, the cosmos, and existence itself. At the Johns Hopkins Center for Psychedelic and Consciousness Research, David Yaden calls them "self-transcendent" experiences, or "those profound moments of connection with something greater than oneself."

Examples of transcendence-triggering activities might be reading a deeply moving book or poem, achieving peak performance in a sport, being in a state of flow while sculpting or painting, attending an impactful mindfulness meditation retreat, feeling gratitude for a selfless act of kindness, merging with a loved one, experiencing awe at a beautiful sunset or the stars above, being so inspired by an idea that you feel "awakened," or having a profound mystical illumination—during a psychedelic experience, for example. As you can see, they occur in many different contexts and at different levels of intensity.

"While transcendent experiences differ in various ways," Kaufman writes, "they all have in common weakening of the boundaries to connectedness with others, the world, and one's own self."

It's that last point—connectedness with one's own self—that distinguishes Kaufman's definition from Yaden's, and especially from mainstream notions of ego dissolution. The core premise of Kaufman's book is that transcendence is not an escape from the self but rather the "healthy integration of one's entire being." A whole self is a connected self. Notions of self-loss, decreased self-salience, and ego death distract us from what's really going on: moments of great connectedness, in which we don't "lose ourselves" so much as become more whole.

In Kaufman's view, the transcendent self is in many ways the healthiest self. Why? Because reaching this point is not a matter of stringing together a bunch of peak experiences and calling yourself "whole" or "enlightened" or "connected." It requires a solid foundation of existence built on safety, connection, self-esteem, exploration, love, and purpose. Once these needs are met, a person can be called "self-actualized," meaning they generally operate from a place of growth rather than deficiency. These are the healthiest individuals, capable of rising to meet their highest potential or true calling. Transcendence is not about the end-fulfillment of that potential so much as the continual "synergy" of it, to borrow a phrase from anthropologist Ruth Benedict, so that when you are pursuing your most selfish gratifications, you are automatically benefiting others, and when you are altruistic, you are automatically rewarding and gratifying yourself.

As Kaufman puts it, "Healthy transcendence involves harnessing all that you are in the service of realizing the best version of yourself so you can help raise the bar for the whole of humanity." When people reach this state, they tend to experience sudden peak moments or longer "plateau" periods more frequently, as well as feelings of flow, connectedness, and unity. They can leverage these moments to create meaning for themselves and others.

"Self-transcendent experiences appear to be potent sources of well-being and prosocial behavior," says Yaden. "We have found that spiritual and self-transcendent experiences are associated with positive outcomes related to mental health. We believe that by helping people to understand that many other people have these experiences, and by helping people to understand their experiences, we can help people to integrate them into their lives in a more positive way."

Yaden, who studies the science of transcendent experiences, was originally inspired by neuroscientist Andrew Newberg, MD, who scanned the brains of expert meditators (from Tibetan monks to Franciscan nuns) and found changes in the superior parietal lobe, a region of the brain associated with spatial body awareness and social cognition. Since various neuroimaging studies have found that temporal, spatial, and social distance may be mediated by similar brain processes, it's clear that feelings of transcendence are more physical than we might imagine.

"During self-transcendent experiences, people may feel deeply connected to other people (their social environment) as well as objects around them (their spatial environment)," Yaden says. "This feeling of unity may result in attributing social qualities to one's spatial environment—a social/ spatial conflation."

He thinks that blurring the lines between the social and the spatial, or perceiving the social in the spatial, may be another way to increase perceived social connection—and thus increase well-being.

Yaden and Newberg's book, *The Varieties of Spiritual Experience* (2022), was still being written at the time of my own research. Having read Scott Barry Kaufman's book, however, I wrote to him in April to let him know it made an impact on me. Here's our social exchange in cyberspace. (You can tell it's week six or so of the pandemic by all the exclamation points.)

SB: I just finished *Transcend* and wanted to thank you for writing such an important book. There are so many useful insights in Parts I and II, and the way you layered each section upon the last helped me see how all of these things—safety, connection, self-esteem, exploration, love, purpose—interact with one another. Part III actually kind of changed my life. Thank you so much for highlighting that we all have a bit of the transcender in us, and that these values should be taken seriously!

SBK: YES!! So glad to hear that, Saga Briggs!!!!!!

SB: Please let me know where I can write a review! Also, I think you'd be interested in Alan Page Fiske's new book, *Kama Muta: Discovering the Connecting Emotion*. He argues that humans have evolved to value peak experiences simply to motivate us to seek connection.

SBK: Good stuff—thanks for bringing it to my attention!! Would you mind writing a review on the Amazon page?

SB: Done! I got a little excited and did a write-up about your book & Alan's on my blog as well. I can't promise anyone other than my mom will read it, but it felt great to write! Thanks for the inspiration.

SBK: Thank you so much. I really appreciate it.

OK I'm definitely going to have to check out Kama Muta.

Also you have the coolest name ever.

OK terrific post. Would you like me to share it?

SB: Thanks! The name's a lot to live up to—no pressure from the parents there. Feel free to share the post! And enjoy researching kama muta. Alan's a really nice guy and would definitely chat with you if you reached out to him.

SBK: Do you reckon he'd make a good guest for my podcast?

Your post was very thoughtful. It was clear you actually read my book. I must say I'm impressed!

SB: Yeah, I think he would! You two would play off each other well. And my pleasure. It's an easy book to recommend.

SBK: Thanks, Saga!! Hope you're staying safe and finding meaning and creativity during these highly uncertain times.

SB: Thanks, Scott! Strangely feeling more creative in some ways than before, so I'll keep riding that out. Wishing you all the same and more.

> C and I first met on the subway on Christmas 2019. None of my close friends were in Berlin, I wasn't talking to my family, and my inner world felt blown apart. But something about it all made me feel stripped of defenses, more human, and when I sat down in the seat across from him, I caught his eye and held it for longer than I usually do with strangers.
>
> "If you want to know how someone feels, just look into their eyes," writes Julie Holland, MD, in her book *Good Chemistry: The Science of Connection from Soul to Psychedelics.* "When we read the eyes this way, the information is processed through the insula, the seat of our intuition."
>
> I saw what he saw in me: something to learn.
>
> We studied each other for a moment, and when a man walked by holding out a paper cup, C reached inside his cartoon fox–patterned tote bag, held out an avocado, and shrugged as if to say, "I can't give you what you want but I can give you what I have."
>
> We rose at the same stop and I told him I liked his bag. "Thanks," he said. "I like your eyes."
>
> The next day, he called me in the middle of the day. I texted back. Soon I discovered he was the type of person to call whenever he wanted to talk or make plans, which gave me mild anxiety at first. But once I got used to it, especially as the pandemic set in, I thought, "This is nice; why do people not call each other anymore?" The obnoxious, hyper sound of my phone ringing—which before that point had stoked my various neuroses—became my favorite sound of 2020 because I knew it had to be him.

NATURE

In December 2019, Alan Page Fiske and his colleagues in Norway released a perspective article on the psychological processes underlying connectedness to nature. They argue that the positive feelings evoked by forest bathing or other nature activities may, in fact, be social emotions. "Social connectedness and connectedness to nature are underpinned by the same emotions," they write. "Social-relational emotions, especially kama muta, seem to be salient in experiences of connection with nature."

Led by researcher Evi Petersen, the team curates findings from ecopsychologists, social psychologists, philosophers, spirituality researchers, humanistic psychologists, and others to highlight the range of relational experiences that can lead to feelings of connectedness.

"Humans are social beings and therefore have a fundamental need to relate. This need is often satisfied by socially connecting to others such as a partner, family, or friends. However, we know that people also socially relate to animals, deceased ancestors, deities, abstract entities such as countries, humanity as a whole, or even imagined collectivities in order to meet their need to relate. Likewise, ecopsychologists have pointed out that the need to relate can be satisfied by feeling connected to nature."

Abraham Maslow's peak experience studies in the 1960s showed that 82 percent of participants had experienced the beauty of nature in a deeply moving way. In 1998, ecopsychologist Jody Davis, PhD, proposed the term *transpersonal experiences in nature*, which includes the experience of peace, joy, love, support, inspiration, and communion. In 2005, Paul Marshall outlined mystical experiences in which people felt that the natural world evokes a sense of unity, knowledge, self-transcendence, eternity, light, and love. More recently, in 2017, Ryan Lumber and colleagues investigated the Biophilia Hypothesis, which links nature connectedness with the evolutionary notion that humans depend on their natural environment, and found that contact with nature, emotion, meaning, compassion, and beauty are pathways to improving nature connectedness. And in 2018, C. L. Anderson and colleagues found that experiencing gratitude and awe while in nature predicted stress reduction and increases in well-being among military veterans and youth in underserved communities.

So what does this mean? If the literature suggests that social emotions like awe and gratitude play a direct role in the well-being that results from contact with nature, then it's plausible that spending more time in nature may make us feel more connected to people. Surely it's not a substitute for the real thing, but it may help us get by when we're feeling the distance.

At the University of South-Eastern Norway, Evi Petersen, PhD, studies what it means to feel connected in nature. After Alan sent me her way, I Zoomed with her to find out more.

SB: Do you have any theories as to why social emotions like kama muta are experienced in nature?

EP: There is no such theory out there which could just explain how we can make sense of these emotions in the outdoors or in connection to nature. But we can build on some general assumptions, and that's what I'm doing with my work. So I'm looking at cross-cultural aspects of nature connectedness to see where the link is between what's very individual and maybe only applies in one cultural setting, but also where can we find maybe more fundamental aspects to this connectedness.

And so I'm going back to thinking about the basic idea of experiences, how they influence our life. That's why we make decisions. That's how we go about life. And this always builds on whether we ascribe something meaning. So we have to go back then to: what is the meaning-making process? And for me, it is a psychological approach that I'm taking where I'm thinking more generally, that if we have a feeling of something that's meaningful for us and makes us feel connected to something, then that itself is a psychological process. So it's not so much about the context, but the process itself as a psychological one, and therefore it could apply to everything. It could apply to [feeling connected to] humans, animals, God. It could apply to everything where humans have the cognitive capacity to make up a concept that they can relate to, which is meaningful to them. So I'm taking it from there. From a fundamentally cultural, psychological approach.

SB: So it's about the meaning you make out of the feeling more than the feeling itself.

EP: Right. If an experience is meaningful to you, it could be meaningful because you have already experienced something else that makes that meaningful to you. But it could also be because you have certain expectations toward it. Otherwise we wouldn't have ideas and concepts like, for example, God, or even being part of a soccer community that doesn't exist on paper, but everybody feels connected. We are just creating something that we can relate ourselves to.

And so the mechanism of relating yourself to either something that is very physical and you can touch it almost, or it's something that is more abstract . . . it builds on the idea that the human brain has the capacity to make up concepts which are meaningful. And if you have that capacity, you can apply it to nature as a concept, as a whole, but you could also apply to just one single element in nature, which could be an animal or something, and it's similar because you don't connect to each element just like you don't connect to every human. You have a meaning-making process behind why you are relating to somebody. And it's the same for nonhuman entities, including parts of nature.

SB: I'm curious which parts of the brain are activated during an experience in nature like that as well. Would you say it's the same parts that are activated when you're interacting with another person?

EP: There are many ways of looking at it, and I'm looking at it from an emotional research perspective. So for me, emotions play a very big part in why you can have an established feeling of connection to something like, for example, nature. If I look at social relational emotions, those emotions seem to trigger exactly the same aspects in the brain for the amygdala [limbic system] as human interaction. In the particular study I saw from Lisa Feldman Barrett, and in my own observational research, people had different emotional triggers, but it didn't make a difference what triggered the emotion, if it was caused by people meeting each other or by having some other sort of interaction with other nonhuman entities. It only made a difference if the person was ascribing meaning to the situation.

SB: So connectedness may just be a construct.

EP: Yeah, it's very much about you and it's very much about your personal interpretation of the situation, which could be very different from person to person.

SB: Would you say that kama muta is an emotion or a mood?

EP: If we talk about emotions, we come from appraisal theory. Because there are many ways of looking at what is an emotion, a feeling, a mood. We're trying to navigate through the landscape by defining an emotion as something that doesn't last very long. It's only a sudden intensification. So you only have that for a maximum of three minutes. It can't last longer. Then we wouldn't say it's an emotion anymore; it would be a mood or something else. Or are we then already describing the relationship which comes out of it? So when we talk about the emotion of kama muta, we always talk about this moment, this situation where the intensification of the relationship is.

SB: One thing I found interesting while reading about kama muta was the interoceptive characteristics that come along with it—warmth in the heart, teary eyes, goosebumps, a lump in the throat—since I'm interested in interoception as a way of understanding emotional experience. There's more and more literature in psychedelic science highlighting the felt experience alongside brain activity—for example, studying MDMA's effects as an entactogen.

EP: Yeah. And I even think that taking MDMA would probably allow more people to experience emotions like kama muta, which can help to intensify communal sharing and connectedness. Or to really establish connectedness, like the feeling and the permanent relationship of connectedness. Yeah, that would be a great study.

SB: What kinds of measures do you use to identify kama muta in study participants?

EP: Valence is one of them—whether you feel pleasant or not. Then we have the verbal construct [interpreted differently according to culture]. And then we have the sensation. There are five categories of sensation and you have to have two of them always. Otherwise we would not categorize it as kama muta, but maybe more something like gratitude or some related emotion, and it's not even clear where to cut the line. Those emotions are quite fluid and they go into each other, and what I see from my data is that people often have many emotions at the same time. And that depends very much on what kind of meaning-making process you are going through. How does this experience make sense to you, how do you make meaning out of it? You are always reinterpreting your situation, so that's why you have quick changes in your emotions and sometimes two at the same time, or three.

SB: Would you be willing to share a personal experience that you've had connecting to nature or experiencing kama muta, and what it felt like for you?

EP: I have very different and connected emotions to different landscapes. One of those landscapes is from my childhood, and I think that's the case for many people. When I think of a good place to be, where I feel comfortable and where I can relate and feel connected, I go back to the waterfront where I grew up and I have those sensuous experiences. I smell, for example, iodine. It's very iodine-rich air where I come from. And we have the mudflat. It's the northern part of Germany. So you have the ocean, the walk on the mudflat when the water is gone; when the tide is low and you have this open view, you can breathe in this very particular smelling air. And as you probably know, smell is one of the senses which are absolute. You don't forget them. Scent is close to the emotional part of the brain, so you can have those emotions right away, and for me, it's elicited when I smell this iodine and I see this ocean and I feel the wind on my skin, and often it brings me to tears because I feel very much connected to the landscape. And it's almost like a symbol for me to relate back to where I come from, to my roots, to my cultural identity. It really brings me a lot of peace, and I feel calm and refreshed. And I have this experience almost all the time when I go back there. That's it for me, one of those particularly important landscapes and emotions that are out there. I would like to know, do you have any experience from being in nature or do you have any personal connecting points there?

SB: Yeah, so I grew up on the Oregon coast in the United States, in the Pacific Northwest. I'm from a really small town of 600 people, and it was all old-growth forests and mountains right next to the ocean. So I feel very, very connected to nature. And it's kind of crazy that I live in a huge city. But Berlin is close to 40 percent parks and forests and lakes, so somehow it's not so bad. But yeah, I often wonder if you do grow up surrounded by nature versus growing up in an urban area, if you end up being more receptive to these kinds of emotional experiences, or if it's, again, kind of the same question of culture and background.

EP: Yeah, there is some research on it, and it is very interesting that people from the city are much more aware of protecting nature. And much more open to nature experiences than people from the countryside. And it is because it's more valuable. Whatever you can't have is more valuable to you. But then [if you are from the countryside] you don't have these very intense emotions of: wow, look at that. It's not surprising you. And if you don't have the surprise or something is changing, there's not much happening on the emotional part of it because your system is in the state of balance.

SB: "Whatever you can't have is more valuable to you" includes connectedness itself more broadly, doesn't it, especially considering this year?

EP: Yes, it's really interesting in the context of [the pandemic]. What does it mean, not being connected in the ways we used to be, and how can we establish other ways? I'm very interested in haptics or the sense of touch, for example, and why it's so important in establishing connectedness. What is going to happen now with COVID, and how many people will suffer from it? We are lacking this connectedness which could be established through physical contact, and I mean, it's probably going to be an extra chapter for your book.

SB: I would say that being connected to nature got me through March and April personally. I live close to one of the biggest forests in Berlin, so I was always out walking around and hiking. And I felt it before I even knew about your research: "Okay, yeah, this almost feels like a good substitute. It's not the real thing. But it feels like a relationship."

EP: I think it is a relationship. It just has different qualities to it.

SB: Right, so it's definitely a time when people are being forced to be creative with how they connect. Being in nature, getting pets, taking psychedelics, living their lives through Zoom . . .

EP: Do you have any experiences yourself with taking psychedelics and really having this feeling of connectedness?

SB: Yeah, I do. I would say mostly with psilocybin, and I've had one experience in nature that was really powerful . . .

EP: Were you combining those aspects, or . . . ?

SB: I've tried on purpose to do it in different settings, so yes. But the set and setting really matter—it has to be in a safe environment where you trust the people you're with and maybe already have a bit of a connection there. And the feeling kind of just . . .

EP: Intensifies?

SB: Well, I was going to use the word *intensifies* or *enhances*, but I kind of back away from that language because if you come at it from a predictive processing or predictive coding point of view, what the compounds may actually be doing is not enhancing empathy or connectedness, but removing a perceptive filter that you already have in place. And so it's almost like these are the feelings that you would normally have if you weren't so desensitized or traumatized or didn't have a negativity bias holding you back. Like, this is actually the [underlying] state. I would say that actually rings true based on my experience.

And then I've definitely had experiences where I've done it alone—like an internal journey with eyes closed and blocking out all external distractions, because there are often visual components to it, and you can choose where to place your attentional resources—and that's been really different and maybe led to more interoceptive experience than it would have otherwise.

EP: Yeah, that's super interesting, and in my research, I see that people use nature as a gateway to those different connecting paths. So they connect to humans, they connect to nature aspects, they also connect to themselves. They use nature as a way inside. And, well, it depends very much on what the need is in the moment and what they are already like, what kind of mindset they are in, I guess, and also what nature the landscape is doing to them. Also how they get approached by nature, for example if there is a bird coming and sitting next to them or not. So it's really important to think about the interaction that comes from nature itself. But I find it interesting that those connecting processes often glide into each other. Having a nature moment with an element of nature, with another animal, can also often trigger a more personal journey with yourself.

Have you experienced something like that as well, that those connecting processes go into each other or were they more separated from each other? Do you see that connectedness can be fluid and you can connect to something else as well, that it overlaps?

SB: I mean, I think so. I don't know if I was aware of it in the moment that way. But I guess it's almost necessary that, in order to feel more connected to others, you have to feel connected to yourself first, or to notice the internal feeling and have some sort of back-and-forth noticing dynamic of your experience and the other person's experience. I don't know if they can even be separated. I'm not sure I have all that much experience with the fluidity like you're talking about in the moment, but I definitely will sometimes have body-based insights or just cognitive insights from an experience in one of those categories that I'll then take the next day or the next week to another situation and try to apply it. So maybe there's a lag time between when things interact.

What about you? Have you had any experiences where kind of all of these things sort of overlapped or integrated together?

EP: What I quite often experience is that when I am alone, when I'm solo—that's what we call it, "solo in Nature"—that I then feel much more connected to anything else than if I'm, for example, in a group. It's probably because you have the opportunity to think. You have the need—because you are alone—to connect, and you also have the possibility to think of constructs which you are linked to and maybe reflect about yourself and your connections in life. I think a lot of people that I talk to value exactly this aspect, being alone in nature, making you feel connected to others or to something. And for me, I often have situations where I have a connection with an element of nature—which could be, for example, just a bird—and I start to project some of my ideas or emotions onto that bird. So, for example, this bird has some intentions regarding me. And if I build up this connection, it often ends up that I reflect more about my overall connectedness in life. So for me, it's often thought processes which start as one single connecting aspect, and then go into bigger reflections about how I'm actually integrated into other parts of my life. But it's interesting for me that it's often triggered by one connection and then kind of triggers some other aspects of me thinking about, "How am I connected?"

But I'm not sure if many others have experienced this as well. I only know that Moreton had a study in 2019 on the relationship between broader nature connectedness and social connectedness—if there is a relationship or not—and he found that the abstract form of nature and social connectedness do have a connection, but not if you think of a single connection. So, for example, if you're feeling more connected to your dog, you're not necessarily more connected to your friends. So there is not a real connection. But if you feel connected to the broader concept of nature, you often also feel more connected to humanity. It's these broader concepts which are more connected to each other.

SB: And it's interesting that you can be in nature with a group and have the opportunity for social connectedness right there, but it can sometimes be more authentic, actually, to be on your own.

EP: It's taking away some barriers. So maybe it's quite similar to the psychedelic experience where you are also taking something away, as you said, to make other things flourish.

SB: One last question—what do you think of the biophilia hypothesis as an explanation for why humans feel connected to nature?

EP: I think it's very beautiful. The problem with this is that it's not a real theory and, as the name says, it should be a hypothesis that is falsifiable, if possible, which it's not. So if you think of it in an academic way, the problem is you cannot falsify it because it will always be true. It always has another loophole for answering why somebody would not feel connected, so that's a problem with it. So I think it's more of an axiom. It's more of a spiritual idea—you either agree with this general idea or you don't. But out of this biophilia, some other theories have evolved, which I find much more solid or much more applicable.

SB: The first thing that comes to mind for me with the biophilia hypothesis is, if it were true, we wouldn't be having these climate change issues right now.

EP: And that's also why it's so beautiful. It's also the answer to that: all of this evolved because we are not connected anymore to nature. You have to be more connected to nature and then we will figure out that that is wrong, what we are doing.

SB: And we'd better do that fast.

EP: Yeah, but I think the biophilia hypothesis comes from the Western world, this idea and the theory. And it doesn't capture that we are culturally very, very diverse. Most of our research is only in the Western countries and we kind of forget about the majority of the population in the world, which is not Western. And there you get a problem of when you want to apply this theory or this hypothesis. The cultural lens is super important for all of us. And I think that's also a movement that we see, at least in psychology—that people are more aware of it, starting to think, "Well, is that actually the case for everybody?" I think it's getting there, and I find that a very good sign.

> Spring came early to Berlin. I bought a new bicycle. Shops opened up just enough to serve ice cream from the storefront windows, to keep the kids happy. Everything worth doing involved being in nature. With one friend at a time, I would kayak or swim or bike to new forested destinations outside the city. But the best nature excursions happened at night. One night in April, C and I met at the edge of Grunewald, the 7,000-acre forest close to our neighborhood in the West, famous for its population of wild boars. We explored for several hours in the absence of moonlight, relying on our other senses to lead us through the dark. It felt a bit like being trapped inside a small room and using your will to expand the walls. The act of straining my eyes

was disorienting until my ears and nose and skin and instincts took over. We seemed to be finding our way. One minute we were cackling about something funny, the next we were still as posts, listening to a rustling and grunting in the brush. Terrified, he started to speak. Terrified, I hushed him. We scanned the nearby trees to see if we could scale them. Slowly and silently, we walked backward the way we'd come, until the trail made a sharp turn. We ran until we reached a road and saw headlights coming toward us, then ducked into the brush and rolled, laughing, onto the leaves and out of sight. These were the most connected nights, and the most memorable: seeing with our bodies.

BODY, EMOTION, ACTION

Kelly Mahler, OTD, is an occupational therapist serving school-aged children and adults. She is actively involved in several research projects pertaining to interoception and has published seven books, one with neuroscientist Bud Craig, PhD, on a variety of topics related to social and emotional growth. I have personally completed Mahler's Interoception Course and can highly recommend it. We caught up a few weeks after my chat with Petersen.

SB: Can you talk about a couple of the practical strategies that you use with kids to help them connect with their bodies?

KM: We use the Interoception Curriculum, which is a systematic step-by-step process that we follow. Step one is "Body," teaching individuals to notice their bodies and also noticing a racing heart or tight muscles or sweaty hands. Then step two is "Connecting Emotion," and it's helping them to know that they're noticing body signals. What do the body signals mean for some people? They have to be systematically taught that, especially for individuals that have autism or ADHD—they might notice a growling feeling in their stomach, but they have no idea what it means. So some individuals need to be taught that. And then finally, step three of the Interoception Curriculum is called "Action," and that's using feel-good actions to promote positive mental health and self-regulation.

So some of the practical things: if we're back at step one—Body—we chunk that work into 15 different body parts so that we're not overwhelming the individual, and they work within just one body part at a time and noticing body signals in that one body part. We start with body parts on the outside of the body because that is very abstract. It feels a little bit more safe. And then we slowly work to notice body signals on the inside. Our very first lesson is on hands and noticing all the different ways your hands can feel. We do different experiments that evoke interoceptive sensations in the hands, like holding an ice cube makes your hands feel cold or putting your hands in water makes your hands feel wet or squeezing a stress ball makes your hands feel tight. And they get practice noticing those body signals in safe and playful ways through these experiments.

And what those experiments are doing is also giving concrete meaning to interoceptive language, because for a lot of our learners, tight hands or a racing heart is so abstract; they don't concretely understand what that means. So we move through the program that way, changing up the body parts. Lesson two is feet and all the different ways your feet can feel. And then we do mouth and eyes and then we move slowly into the inside of the body. Then there's a lesson on the heart, all the different ways your heart can feel, then lungs, stomach, bladder, and so forth. So we chunk it that way. And we've heard even from the brightest adults that have gone through our program—they love how slow it is because it can be very overwhelming for a variety of reasons to pay attention to your body.

SB: Is there a strategy you have for helping different individuals notice the nuances between their own feelings?

KM: Absolutely. That's why we move into step two of the curriculum and that's the emotion piece. It's teaching them to use their body signals as clues to their emotions and helping them, in a really individualized way, discover, "Okay, when I notice these body signals, I essentially need to add them together, and what emotion does it mean for me?" Because what your body feels like when you're anxious is different from what my body feels like when I'm anxious. But a lot of the curriculum and the mental health strategies out there teach a one-size-fits-all approach, like, anxiety looks like this: speeding heart, sweating palms. And that's not true for everybody. So we're really trying to find an individualized approach and help people just be more curious. There's no right or wrong answers in your interoception journey and what you're discovering. So we really try to be clear on that, that it's so different for all of us.

SB: And can you talk a little bit about why you think autistic individuals tend to have this issue with interoception to begin with? Do you have any theories on why the two are related?

KM: I'm not sure I'm ready to be quoted on any of these theories yet, but . . . I mean, I don't know if it's just differences in their neurology from day one or if it is their differences in neurology in other areas that cause this. For example, if they're born into a world where they are super sensory-sensitive to external sensory information, their attention is obviously going to be pulled outward into the environment in a hypervigilant state, in a protective mode. And so they have fewer resources left to pay attention to their inner sensory experience. But that might not be the case for everyone. And we know that not all autistic people are sensory-sensitive to the outside world. They might be the opposite. They might be hyposensitive. So I think that's probably only for a set of individuals.

SB: Do you have thoughts on how interoception might be related to some of the therapeutic effects of psychedelics and entactogens, like MDMA, for example, to treat trauma?

KM: It makes sense to me, in my understanding of trauma, that many people become completely dissociated from their body out of survival. So if you have an infant from a young age and their interoceptive needs are not met and not validated, they feel hungry, they cry, their caregiver doesn't attend to them. So they begin to learn that "My body skills are not important," or they disassociate from their body because their body feels so intensely uncomfortable and overwhelming because their needs are not met. So then maybe the MDMA experiences open them back up to that window of "Are my body signals important?" And if they're being validated in that moment, then maybe they can start to rewire their brain and almost come back into their body, if you will. I mean, that's just a guess, but it makes sense to me. And what we're seeing clinically—it's a lot slower of a process because we're not using psychedelics and we're working with young children. But it makes sense because we're all about bringing them back into their body in safe and playful ways and validating their inner experience, and that has been very helpful.

SB: What made you get into interoception? How did you discover that this was an important area to study?

KM: I was writing a book with a friend a few years ago, and at that point I had only heard the term *interoception* and thought it was just a toileting thing. I didn't have any idea what it was, to the depths that it is. But through writing that book—it was a book on autism and sensory processing—I did a lot of research and started diving into this eighth sense that we were adding to the book. And I realized through all that research, "Oh my gosh, this is such a big deal." So I ran right to my clients and their families and started asking lots of questions. And we discovered that nobody was talking about it and nobody is addressing their interoception needs. And it was influencing all the goal areas that I had. And we were getting certain gains on their goals, but we weren't really getting the fullest extent of what we could until we started incorporating interoception. It was a game changer. And still, there's not enough people that understand it and how impactful it is on all of our lives and so many different areas of life.

So that's really my mission, to educate people about the topic and get more people on board and really help disseminate this into the practical application space, because neuroscience, they got it down. You know, psychology and the research, they're getting it down, but nobody is really shifting it to the practical application side of things and it needs to happen now.

SB: Yes. Shifting it into an embodied space.

KM: I know. For so long, our whole mental health field got so cognitive and brain-based that we forgot there's a whole body down here.

SB: Yes. There's Somatic Experiencing and all of this body-based therapy now, but it's also still very focused on the individual: you practice feeling your own body and it's not really social. Yes, there's a social dynamic between a patient and a psychotherapist, but there's nothing really that I've seen where the format is sort of like embodied group social therapy or something, where you're in touch with your body, amongst other people, which I know gets kind of tricky, especially if it's a trauma patient or something...

KM: You're making me really think about that a lot, because last year we did a big study where we had over 100 kids enrolled in using the interoception curriculum and they were in different sites, some were summer school–based and some were in outpatient. And the school-based kids—just an overall glance of our data—did much better and their outcomes were much stronger (their ability to emotionally regulate after eight weeks of our program versus the outpatient care). And I'm wondering, just based on what you're saying—the kids in school were getting it in a group, in a classroom setting. So they were all able to hear each other's experiences and they're kind of connecting over that. And even the kids that didn't want to participate at first—we never force anyone to participate— they saw all the other kids doing experience, experiment, and noticing the way their bodies are feeling. And they all joined in. We've never had a single participant flat out refuse to participate over the four years of our study. So I don't know. Just makes me wonder, did they have better outcomes because they were inpatient versus outpatient or with a group versus just the outpatient therapist and the child, with, of course, the caregiver—I guess that is kind of a group, but maybe not as powerful as your peer group. I don't know. Interesting.

SB: Yeah, it's interesting to consider. It seems like any kind of disorder or illness—if it comes from lack of sufficient care or social interaction to begin with or, you know, develops relationally, it makes sense that the treatment for it should be relational as well.

KM: Yeah, I don't know if this is your area of interest, but we're doing a lot of work now in early intervention—so from birth to age three—and finding that a lot of the caregivers themselves have limited interoceptive awareness. And how does that then impact their attachment and attunement process? Because if you're not in tune—we know from research—if you're not in tune with the way your body feels, that definitely impacts your connection and understanding of the way someone else feels. So how does that play out in the caregiver–infant relationship? I think that's a really interesting place to look. This is not a research project; this is just stuff we're playing around with clinically right now in order to come up with a good research design. I don't know the way it will look. We have a lot of clinicians out there that are working on doing some really cool stuff with caregivers. And then hopefully the outcome of that would be that it'll trickle down to their children.

SB: I'm so excited about all the stuff you're doing.

KM: I'm excited about your book. We all have to work together. My one main goal is that when I'm texting, my phone will recognize the word *interoception*, or when you're emailing someone, it won't keep changing it to *reception*. So we all need to work together to get it accepted by spell-check. I feel like we'll have made it then.

> Back in Seattle in 2019, my friend and I had gone to see Malcolm Gladwell speak at Benaroya Hall. Something he said during the Q&A

stuck with me: "Drunkenness is not disinhibition; drunkenness is myopia."

Disinhibition theory suggests that, while drinking, you are increasingly insensitive to your environment, relaxing and loosening up as you sip your pint. Myopia theory states the opposite: you are increasingly *sensitive* to your environment, completely at the mercy of whatever is in front of you, whether it's a human relationship or a Jell-O shot.

According to cognitive scientists Claude Steele, PhD, and Robert Joseph, PhD, who believe we've misread alcohol's effect on the brain, alcohol's principal effect is to "narrow our emotional and mental field of vision," causing "a state of shortsightedness in which superficially understood, immediate aspects of experience have a disproportionate influence on behavior and emotion."

Booze makes the thing in the foreground swell with importance and the things in the background dissolve. Salience, disintegrated.

In Berlin, C and I were both recovering from myopia in our own ways.

"Let's get to know each other's bodies first," he would say, suddenly, every time we started to do just that in my narrow bed. Neither of us knew what that meant, perhaps because we had only recently rediscovered our own bodies (he had also stopped drinking in the past few months), but it seemed like a good thing, like reclaiming something: intimacy should create more space for things, not less.

One day, in late spring, I told C in a text that I'd spoken to my mom on the phone the night before. She and I hadn't spoken on the phone for months. A few minutes later, he called me to ask how it went. In the space between our words, the inside of my body started to glow, expanding to make room for him. I wanted to name the feeling, say it aloud, but didn't.

ATTACHMENT

Wanting to dig into Mahler's point about caregiving and interoception, a few weeks later I reached out to Kristina Oldroyd again. We talked about the developmental origins of interoception as it relates to attachment and emotion regulation.

SB: I think you are one of the only people out there researching the intersection of these things: interoception, attachment, and childhood development.

KO: It's interesting how this line of research came to be for me. I was in grad school, and I was taking an attachment seminar and a cognitive development seminar, and doing a lot of work on psychophysiology. And because I had the classes back to back, it was like, week after week, I was hearing the same parts of the brain mentioned. But they weren't mentioned in the same context. Like, nobody was saying that attachment was responsible for or related to or in any way connected to these other parts of the brain. It was like, "Does anybody get that these are the same parts of the brain that we're talking about?" And so I started doing lit searches and nobody had looked at that. So we took extant data that we had in the lab—because, you know, it's hard to get funding for new thinking—and I started going through lab data and talking to professors and saying, "What data do we have that would allow us to look at this even in an exploratory kind of way?" And we had a couple things that we thought we could make work. They weren't perfect measures of interoception and attachment, but we could at least make a case and start this exploratory work.

SB: Can you describe your study on emotional regulation and interoception in children?

KO: So for that one, we were studying psychophysiology of narration, and we had kids come in and tell us about a time that they were angry, and we had them hooked up to respiration, heart rate, skin, electrodermal, and analysis of skin perspiration on the fingertips. And we asked them how angry they were feeling as they told us this story. And what we found is that their psychophysiology to their body readings did not correlate at all in most instances with children's self-report ratings of how angry they were. Their heart rate would go way up, their breathing would increase, or their fingertips would get really sweaty. And the kids would be like, "Oh, no, I'm fine. I'm not angry." And we're like, "Yes, you are!" So we thought, why are some kids really disconnected? Is it that older kids are more in tune? Is it that younger kids are more in tune? What is going on? Where is this disconnect from the body coming from?

SB: Were you able to confirm that these measures were specifically markers of anger rather than a different emotion, like anxiety?

KO: While we cannot definitively determine that what we were seeing was anger rather than generic arousal or another specific emotion, we did several things to try and increase our ability to discern this. First, during elicitation of the memory that participants narrated, we specifically asked people to think about a time that they were angry. Second, we obtained self-report data asking participants to rate on a scale of 1–5 how much they were feeling six different emotions: anger, sadness, happiness, fear, guilt, and shame. We statistically controlled for any ratings on the other emotions. Finally, we obtained three baselines to allow participants to relax from the commotion of the physiological hookup and anxiety of being in the lab. We used individually calculated mean of the difference scores and standard deviations in our analyses. In other words, each person was compared to their own readings at baseline rather than differences from the group mean. This should have controlled for individual arousal at baseline.

SB: And what happened next?

KO: This is when we started looking at attachment and interoception. We had mothers in the other room during the study, just filling out some general parenting measures. And, you know, it would have been great if we had an attachment measure, but we didn't. We did have the acceptance and rejection of negative emotion measure, though. And so that's where that came from, you know, just finding that the more moms were able to deal with negative emotion and to help children cope with negative emotion, the better their kids were at recognizing and responding to their internal feelings of anger or emotional distress.

SB: And this was mothers' self-reports of their subjective opinion on how well they were helping kids regulate?

KO: Some of the questions were things like, "When children cry, they're being manipulative," you know, trying to address the underlying feelings, or "I want my children to feel negative emotion," or "I'm afraid when my child displays negative emotions." Not how moms handled it, but moms' comfort level with the negative emotion.

SB: And that can lead to poor interoceptive accuracy as well as avoidant attachment in kids, is that correct?

KO: It can go both ways. It can make it kind of a blunted interoceptive accuracy or overreactive, depending on the type of attachment. You can see the anxious attachment style leading people to overreact. And I think you see in the literature a lot that maladaptive high interoception is associated with anxiety. You know, people who feel their heart start to beat fast and they're like, "Oh, my gosh, I'm dying, I'm going to have a heart attack any minute." It's like, "No, you're walking up a hill. You're fine."

SB: Would you say it's more about paying too much attention or misinterpreting the signals, or a bit of both?

KO: I would say both. There's an interpretation piece there. I'm really interested in athletes and sports. Athletes get praised for pushing through the pain or playing through an injury. And I've done it. Like, you know, I've had my son get tackled on the football field, and said, "Shake it off, get up." And he played a second half and we took him to the emergency room and he had a broken arm and gosh, you know, you feel like a terrible parent. So there's this line between wanting to validate what they're going through and teaching them to listen. I think there's a lot to learn about how we scaffold and teach kids to listen to what their bodies are doing.

SB: I guess it seems to come down to awareness of the context, like even with extreme sports and professional athletes, maybe that's sort of what you're supposed to be doing in that situation, but you don't want it to transfer to everyday life—ignoring your feelings or blunting emotional pain in any other part of life.

KO: But I don't know how it couldn't. I mean, if you're telling a child to push through physical pain, you know, then I don't know that they can make the distinction. I don't know that our body can make the distinction. We know that taking NSAIDs like Tylenol and ibuprofen can decrease emotional pain, so I don't know that our body knows the difference.

SB: This makes me think about the stress response system (HPA axis) and how some of us are taught to numb emotions or push through pain in times of stress. Maybe better interoception could help with that.

KO: The HPA axis for emotional responding and the interoceptive axis use the same kind of, what I call, hardware. Interoception and emotion is the software—how we're interpreting what's going on with the hardware at the time. We know a lot of research has been done on attachment style and HPA development, so we know that different kinds of parenting create different HPA axis responding. So we're changing the hardware with attachment style of parenting and then interoception teaches us how to respond or interpret that difference. So, you know, depending on the type of attachment you form to a parent, it gives you either a larger or smaller signal to work with when trying to interpret what's going on. A blunted HPA response gives you a lower volume signal to find out what's going on with your body. So these would be the kids, as I talked about in my study, whose heart rate and respiration would jump from a 2 to a 5 instead of a 2 to a 10. They have this blunted response and kind of think, "Well, am I angry?" The kid who jumps from a 2 to a 10 is like, "Yeah, I'm mad," but a kid who jumps from a 2 to a 20 is probably over-responding.

SB: And either way, it can lead to further stress, most likely, right? Because you realize how you feel eventually, but you also realize there's been a discrepancy between what your body is feeling and your awareness of it. And it's taken a minute to catch up. So I imagine how that could lead to stress in a lot of situations, like with people who are overworked or at the office all day and maybe aren't really paying attention to their stress levels or their internal body state moment to moment, and it just kind of hits them later on.

KO: Yeah, I think even simpler than emotional awareness . . . I mean, if we're taking it way back to basic interoception . . . I'm sure you have colleagues—or maybe you're one of these people or have people in your family—who will work and work and work and forget to eat. I have a friend who does this, which is actually life-threatening. She's a type-1 diabetic. And so she'll go along and she gets so absorbed with her mental world that she doesn't feel her body getting hungry or thirsty, you know, and then all of a sudden, for her, it's a crisis. Now, we're not all in that situation where it leads to a hypoglycemic episode. But certainly we know people who will forget to eat. There's a great paper, I think Jennifer Murphy in 2016, where she talks about interoception as the basis of psychopathology and talks about it with eating disorders . . . you know, that with eating disorders you are turning that interoception way off in order to deal with the hunger.

SB: It's like creating a different hierarchy of interoceptive signals.

KO: I think part of interoception and caregiving is teaching kids to focus on which of those signals are important. Like, you know, is it more important that the clock says it's 3:00 or more important that you feel hungry? Which one of those is going to prompt you to eat—your body or the external stimulus? We tell them, "You can eat from 11:30 to 11:45" and then, "You can't eat again until 3:00." But then you look at the eating literature and it's pushing intuitive eating. But we don't teach our kids to eat intuitively. We teach them to eat on a schedule. And so there's this disconnect between wanting kids to listen to their body, and we know that's important, but we can't always do that either. A function even more basic than eating is going to the bathroom, right? Like, infants can go whenever they want and then we potty train them. And it's like you have to just go in this place, and then they get to school and it's now you can only go in this place at this time. Or a kid says, "I have to go to the bathroom," and the parent says, "No, you don't. You just went." Well, if they can't be trusted to know when they have to go to the bathroom and when they're hungry or when they're thirsty, how are they supposed to be able to know when they're sad or when they're agitated or when they're angry?

SB: And do you think there's a relationship between the way children are raised in that way and their ability to relate to others? Do you think that it has an impact on socializing or interpersonal relationships?

KO: So I'm going to tell you about another study and you're going to think I've gone way off topic, but I promise I'll circle back. I had college students come into the lab and play an obstacle course video game. And after they played for 20 minutes, I asked them, what did you spend the last 20 minutes doing? And about 25 percent of my participants said, "Oh, I ran and I jumped and I swam and I climbed this wall." They narrated in the first person. And then I had them fill out about 20 minutes' worth of questionnaires during which time they had access to snack food. And those people who narrated in the first person—who said, "I ran, I jumped, I swam"—ate four times as many calories as the people who narrated in the third person, the people who said "the avatar ran" or "I made the video character swim."

So the people who kept a distinct line between themselves and the avatar didn't eat as much. Whereas the people who really related to the character on the screen and personalized it—their body felt a need to compensate for these calories that they just burned. And so now we're studying why. And what I think we're going to see is that these people who personalized the avatar experience are going to have more activity in the mirror neurons system and in this interoceptive system, while playing the video game, than people who are able to keep that line.

So bringing that back to your question: if you're more interoceptively aware, some people respond bodily to other people's actions more than others. I mean, I can't watch violent TV shows. Other people can sit and watch people get tortured and blown up. Like, *Game of Thrones* for me is a no; I can't do it. And other people are not bothered by that, you know. And so I think there's some work starting to tell us that some of us feel physically what we see other people doing and experiencing, and other people don't. And, you know, that's a really new line of research. It's a great question.

SB: I wonder if that's related to attachment too. I mean, you'd think that maybe someone who's avoidantly attached would not want to be watching anything emotional or anything super visceral, like a horror movie or something. But actually maybe they'd have an easier time because they're not connected. . . .

KO: Yeah, you could really make an argument either way, couldn't you? It would be interesting to see what the data say. So, does interoception help you connect with others? I think on some level it does. You hear a lot of mothers explain that when they held their baby, they felt it physically. They have a physical reaction. And if you can't label that or recognize that, would it affect your relationships? I think it would. So much of falling in love we describe physically.

SB: One more thing. I think maybe I mentioned that when I emailed you a while ago, but there's been some research in the psychedelic science area about interoception in the context of psychedelic and MDMA therapy. Because MDMA is an entactogen, so it's very much a bodily kind of phenomenon as well as having psychological effects. One theory is that a drug like MDMA is able to reopen this period of development of the bodily self, so that you can rewire it. Does that make sense to you at all?

KO: Absolutely. I think that's really fascinating. I don't know a lot about psychedelics. I do know at the University of Utah at the VA hospital, they're using MDMA to help veterans process traumatic experiences. And I think that could be tied to the interoception literature, in the sense that when they're under the influence of this psychedelic drug, they're able to reprocess the physical and bodily signals that they experienced as traumatic in a safer way. But it would be fascinating if that reopening is what's happening. Because once your HPA axis reactivity is set, you can learn to interpret it differently, but wouldn't it be cool if we could go back and kind of reset it or recalibrate it so that it wasn't this constant interpretation? Because I think that when you're relying on interpretation—when you get tired or when you get stressed, those coping mechanisms fall apart. We can't keep that level up all the time. And so if instead of having this overreaction and having to constantly interpret it, it'd be really cool if you could just reset it.

> From June to September, as the world opened back up for a brief window of time in Berlin, the sodden gale of a breakup blew through my body. Every day for the first week, I woke into what seemed like a nightmare. It felt as though a rare exotic flower had been growing inside me and someone hacked the top of it off with safety scissors. Writing these words now, I can hardly connect to that feeling anymore, but I remember the words I put to the experience at the time. As the months went on, I started wondering how much of the suffering was about this person, C, and how much of it was about me. I seemed to be mourning the loss of something older than the relationship itself; the trauma of it didn't quite match up with its value, though I'd certainly felt love growing in my body in the last month or two of the six months we'd been dating. In August I read more about attachment theory, since information makes a great flotation device when there's nothing else to hold onto. In typical fashion, some clarity came not when I was reading research articles and blog posts but rather picking up a random book at a friend's place and stumbling across a quote from New York–based writer Brian Kuan Wood: "Love's joy is not to be found in fulfillment, but in recognition: Even though I can never return what was taken away from you, I may be the only person alive who knows what it is."

SOCIAL INTEROCEPTION

> *"If I have relatively better interoception than some others, the boundary between myself and others is more distinct."*
> *—Andy Arnold*

Part of being authentic with yourself means paying attention to your bodily responses around other people: your gut feelings, heart rate, body posture, facial activity, general feeling of discomfort or ease. If you can notice these things within yourself, you can decide what to do about them, whether it's making a conscious decision to share your inner experience with the other person, taking a moment to pause and breathe, or changing the dynamic of the interaction by shifting the tone or suggesting a different activity.

It takes practice not to be distracted from these internal cues by external factors. Karen Dobkins says:

> One can be daydreaming (e.g., distracted by pondering dinner plans), or one may be trying to "figure out" the other person—what they are thinking, their intentions, what their impression is of oneself. This type of external hyper-focus may take form as fear of negative evaluation or implicit hypervigilance for social threat, found in loneliness. Both the first (distraction) and second (overthinking) type of external attention are roughly synonymous with not "being present," and like the contemplative practices that refer to this as suffering (e.g., Buddhism/meditation), we suggest this behavior is particularly disadvantageous in the realm of social connection.

At the same time, an essential part of quality social interaction is focusing on the other person. So Arnold and Dobkins don't recommend being so caught up in your internal experience that you're ignoring what the other person is saying, or so distracted by the other person that you can't sense your own internal cues.

In a 2019 paper linking interoception and social connection for the first time—and highlighting how loneliness is likely underpinned by specific interoceptive dysregulation—Arnold, Winkielman, and Dobkins explain how learning to shift one's attention might help promote greater connection. Quality connection lies in a nimble balance between interoceptive and exteroceptive ability. They call this balance "flexible switching":

> High-quality social connection is acquired and maintained by the flexible use of interoceptive signals during social interaction, which we have previously referred to as an aspect of "social interoception." . . . The balance is about flexible alternation between conscious external attention [paying attention to the other person] and interoception, so that the internal signals can be appropriately linked with the external information.

As an example, Dobkins says that when conveying one's inner experience to another, it helps to name the objective event followed by how you felt: "When you said, 'This tuna casserole is over-spiced,' I felt hurt and embarrassed."

Interoceptive awareness is necessary, but not sufficient, for high quality social connection, especially in socially challenging situations. "Awareness needs to switch between internal and external experience, in a moment-by-moment fashion," they write, "so that the two may be *integrated* for adaptive learning in social situations."

In other words, the better you get at switching, the more quickly you learn what internal and external cues mean in relation to each other, which can help you manage your relationships.

Good social interoception not only promotes quality social connection, but it can also bring us peace if the social interaction doesn't go the way we hoped: "That is, one can encounter a difficult social situation, and still—through the process of integration—walk away feeling at peace."

Studies suggest that people higher in interoceptive accuracy can better identify negative physiological responses as resulting from "an objective, external social situation, rather than an attribute of oneself."

This is called reappraisal. The researchers explain, "A difficult social situation that activates interoceptive processes (e.g., noticing an increase in heart rate, flushing) can be reframed as a 'challenge' rather than a 'threat' (e.g., 'my heart is beating fast because I really care about having a good relationship with Jane' rather than 'my heart is beating fast because I know Jane doesn't like me')."

Andy Arnold, PhD, is an interoception researcher affiliated with the University of California, San Diego, and Royal Holloway, University of London. He has research experience in behavioral and molecular neuroscience and cognitive and social psychology. As detailed on his website, this included completing an MA in Psychology at the University of Chicago with Dr. John Cacioppo in 2011, as well as stints working at the Kinsey Institute for Research in Sex, Gender, and Reproduction and the Salk Institute for Biological Studies. Andy joined the Winkielman lab in 2013 and has a broad range of research interests in the social-biological realm, including empathy, loneliness, embodiment, mimicry, and interoception. I spoke to him in the fall of 2020 to learn more about using interoception in a social context.

SB: Can you talk about the work on interoception you're currently doing?

AA: One relevant project I've been working on lately is with an honors student at UCSD. It's kind of like an outgrowth of some of the earlier work that I think I talked a little bit about with you last time. This is the stuff on body trusting and loneliness and interoceptive sensibility, which is how much you actually trust your bodily signals and feel at home in your body. That seems to be highly consistently correlated with loneliness such that, the more people are lonely, the less they trust their body. And so the project we're working on now that I'm helping him design is basically seeing if that sense of not trusting your bodily self and just yourself and your judgments overall, whether that also bleeds off into not trusting other people and potentially moderates that sense of loneliness. So if I don't trust my own signals, am I also less likely to trust other people in my world? And could that then lead to further loneliness?

SB: So interoception seems to play a really important role in social life. Have you looked at ways people can improve their interoception, for example, to manage loneliness?

AA: Some researchers are starting to do longitudinal intervention studies with mindfulness and yoga and body awareness techniques from what I've seen so far. First of all, aspects of mindfulness meditation have definitely been shown to increase both subjective and objective indices of interoception. So that's a promising avenue. [In one study] they had people specifically focus on physiological awareness and acceptance rather than just, "Oh, I accept myself." What they found is that those in the condition of physiological focus in mindfulness actually improve more than the general focus of mindfulness. So I think that that suggests that there are ways that we can create new training programs to specifically target interoception.

SB: I think you said you're also a Reiki practitioner, like me, or have trained in the past. Possibly there's an even bigger interoceptive component to Reiki than meditation?

AA: I do think there's definitely a big interoceptive component to it, and especially pushing that envelope of the differentiation between the self and the other is really interesting with Reiki. The energy is flowing.

SB: Yes. I've had experiences of doing it on others where it felt almost like a psychedelic trip. Like, afterwards, the feeling of connectedness to the person was remarkably similar to the effects of MDMA.

AA: I would agree. Even more so than getting a massage. I mean, I've actually not received Reiki that many times, just a couple of times. But I remember my first time. First of all, there's no touching and I didn't really know the person very well. She was actually my yoga teacher at the time, so I knew her a little bit. And, you know, I trusted her. But after the session, my first Reiki session, I did feel much more open and relaxed and sort of connected with her, even more so than actual physical touch through massage. So I think there's something interesting there.

SB: That might be a good bridge to talking about the insular cortex a bit. Because you could talk all day about interoception and not talk about social cognition, but there's clearly a link. I'm kind of obsessed with the insular cortex, but I also know that no neuroscientist would say that one part of the brain is responsible for any particular process. So I'm curious where you stand as far as how important the insular cortex is to interoception in a social context?

AA: I kind of think of [the insula] as the canonical afferent cortical pathway first. If my insula did not have inputs coming into it from my body, I would probably feel less a sense of self because it does seem like on an unconscious, basic level, we derive our sense of self from the integration of those signals. So first of all, if I have relatively better interoception than some others, the boundary between myself and others is more distinct. I have greater self–other distinction.

I guess there are two ways to look at it from that point of view. First of all, let's talk about the process of empathy. For example, when you're really connecting with someone and the energy is flowing, so to speak, but it's still based at least partly on exteroceptive cues like being able to look someone in the face and sense their micro expressions with their emotion, as well as to remember with that self–other distinction that, you know, your pain is not my pain if I'm empathizing with you when you're sad. But I have a prosocial motivation to help you. So keeping that distinct is important because if I did not have that distinction and that awareness of the self and the other while I was empathizing, I might just start losing it and bawling as well and then not be able to instrumentally help you.

And I guess the other thing I would emphasize is that social cognition, generally, seems to happen very quickly and implicitly. We're often not aware of it until we already have a feeling of, "This is a good person or a bad person. I care about this person or not." So I guess I would also tie that into the body-trusting literature. In my insula, if I'm feeling sufficiently distinct, like myself and an emotional person and a social actor in the world, what am I feeling? What are those signals, do I trust those signals or not? Now, if I trust the signals . . . then if you're my friend and I see you crying, I might have a sinking gut feeling like, "Oh, you're sad, I care about that. Let's see what's going on with you." But again, if I don't trust my own signals, then I might mislabel that or I might say, "Oh, I must be hungry. Let me go get a burrito. You will be fine." But to the extent that you appraise your signals in a trustworthy way and can translate that then to adaptive behavior, I think that can be useful for social cognition in terms of the insula itself.

SB: It's funny that better interoception is related to a stronger self–other distinction, because then you have this whole dialogue around the experience of connectedness and what that feels like in a psychedelic context, for instance. And it's like the opposite. You just want to hug everyone. There are no barriers.

AA: Yeah, like the implicit desire for unity is fulfilled in a way—at least it feels that way in the psychological sense. Yeah, that's funny. I can say that some of my most illuminating and healthy psychedelic experiences were based around, maybe simultaneous expansion of my "self" in that I was hyper-aware of my surroundings, hearing things very far off from a camping site, and also emotional vibes from others, yet I was "detached" enough to not care much beyond just doing the right thing (like fending off the racoons that were closing in on our campsite). My "self" or ego was there but the only charged thought was simple action, rather than social comparison in any way.

SB: How would you say interoception impacts emotion regulation?

AA: First of all, it makes me think a bit about interpersonal emotion regulation. I still like the framework of the intra- and interpersonal interoceptive processes. If I'm feeling really down or happy, you know, how does that relate to the social world and how do I regulate it? Do I go out and seek social support when I'm feeling really down? That can be adaptive, and that's one way to regulate my emotions—by using the social world. If I am feeling, let's say, really down and I realize that, maybe I need some help regulating my emotions. It seems more natural to try to regulate it myself first, right? Like, any time I get upset or sad, I don't want to go running to my mom. I need to take some responsibility here. But if I get to the point where I do need some help, then I guess I think of it as almost "nested," you know, like I'll do what I can to regulate myself before I go to others.

SB: That makes a lot of sense. I mean, just from my own experience, since I started reading more about interoception, I've tried to notice it in myself and I had one experience where I felt angry, what I thought was anger. And then I took a closer look and realized that the feeling was actually happening in my stomach and in my throat. That's where I felt the emotion. And then I realized that actually I just had something that I really needed to say to someone that I'd been holding in for a long time. My body was actually telling me what to do. So I can see how that might help, paying attention to where the feeling is happening and tying that to adaptive social behavior. I mean, emotions happen so quickly. You just want to know what you're experiencing. So it's so easy to just pass it off as anger and then all of a sudden you're making a decision you regret later.

AA: To offer an example that gets at it from my own life, I actually went through a pretty surprising and intense breakup earlier this year where basically, you know, I was with this person for a long time and we were long distance for a bit, came together, and it seemed like everything was going really well. And then I started seeing some behavioral changes in her and I wasn't sure where they were coming from. And I basically had this—well, maybe we could call it a sunk-cost fallacy as well. But I had a strong motivation to still want to believe in her and in the relationship, you know, and sit down and empathize and talk and work it out. But I was getting signals at this point from my body that were just like, no, get out of there, leave. And my body was right. So sometimes you need to override these beliefs as well, and having better interoception can, I think, help you do that.

SB: Yeah, I think that makes a lot of sense. And we live in a world that's so focused on thought patterns and overanalyzing everything, right? You can see why this could help with depression, for example, with rumination. Even in the scenario of a hangover from a bad breakup, I mean, you might have a visceral experience of it for a while, and then eventually, it's like, maybe you think you're not over it, but it's just your mind going around and around and actually your body's fine. I like this idea of listening to your body first and seeing if that matches up with what's going on in your head.

AA: I like that too. Yeah, I think that's a really good point because, I mean, I'll just come right out and say it: it was a very traumatic experience for me.

You know, there was physical and emotional abuse going on all of a sudden that I didn't recognize. I came away from it, you know, mostly physically unscathed. But there was trauma that I really experienced there. You know, I've dealt with some PTSD and some gnarly stuff after that to sort of recover. And I think because I've always had this interoceptive focus, I just dove deep into it. I didn't really compartmentalize that much and try to go about my life because I realized that oftentimes when people do that, the trauma gets stuck in the body, so to speak, and it can probably attenuate the gain on your interoceptive processes as well. If there's something in my body that was just traumatized, but I don't want to deal with it, I'm just pushing it down, that might lead to less trusting of my body and my signals, and you get into the cycle of being even more in your mind and just ruminating—"Oh my God, what happened? Like, who was this person? How could I make this mistake, blah, blah, blah," rather than, you know, digging deep. But I think in some ways after some time of doing that, just like you said, the source of the malaise from the trauma in my body was more cleaned up than my mind expected, but then once I reconnected them, I was like, "Oh, well, I guess time does help and, you know, I'm okay now."

SB: Thanks for sharing that. I also had a rough breakup earlier this year. So that's why I have been thinking about this as well.

AA: Oh, sorry to hear that. Must be the year for that kind of thing.

SB: Thanks, yeah. But you know, once you begin paying attention, it's like this entire other world opens up, and you really can use interoception, experiment with it in different scenarios and big and small happenings in your life.

AA: Yeah, I think so, too. It also reminds me of—I don't know if you've ever read Eckhart Tolle. *The Power of Now*, you know that book? I've been listening to some of it lately, and of course it's sort of similar to other ideas and spiritual literature. But to the extent that you really believe that you create your own world and that you are the eyes of the world, to paraphrase the Grateful Dead—just really trusting that from the inside, you know, grounded in your body—I think you have a stronger ability to recover and ultimately function adaptively than the other way around.

> One night in September I took a high dose of psilocybin and closed my eyes. For a long time it was just a lot of visual fireworks, rapid-fire

> and kaleidoscopic, in varying dimensions and depths, behind my eyelids. After a while I thought, "What's going on here exactly?" and I actually said aloud, sitting up and speaking to the altar at the foot of the bookshelf, "How do I know you're here?" meaning the Spirit as opposed to just my brain doing psychedelic acrobatics. And I felt a little afraid, but also brave, and lay back down. The fireworks continued and I started thinking it was beautiful and impressive but not necessary. Like the Mushroom was just trying to show off. I said, "You don't have to do that. You don't have to try to impress me." My orientation toward it crystallized, so that it seemed I was the Teacher, and the Mushroom quite childlike and adolescent, and I said, "You're not like I thought you'd be," meaning an authority. I felt the relationship was actually me teaching the Mushroom, very gently, like it was a child. And that's when the image of a wound appeared. Blood and something hard to identify, but disturbing. I said with compassion, "You don't have to show me that." Like speaking very gently to a sweet child. *You don't have to show me that.* And that's when the dynamic shifted abruptly and the Mushroom took charge and plunged me into the body of my child self. I have never felt so close to myself, never experienced such profound self-compassion, as I did during that trip. The Mushroom knew that was the only way I would be able to access that level of emotion for myself—by making me think I was feeling it for someone else. This is the bodily lens it was telling me to live through, every day, if I could. After the trip, the breakup from that summer left my body.

EMPATHY AND COMPASSION

> *"Compassion is not about being nice. Compassion is realizing that as soon as you start to judge somebody, whatever adjective you have assigned to that person, you are that thing too. Then instead of separating, you connect." —Karen Dobkins*

In her research at the University of Zurich, Katrin Preller has found emotional empathy to be one of the signatures of positive psychedelic experiences. "We see an increase in emotional empathy, which may be an important factor contributing to the feeling of connectedness," she told me. "In clinical trials, we are currently testing the hypothesis that this experience contributes to the efficacy of psychedelic-assisted therapy."

Emotional empathy is one type of empathy, often contrasted with cognitive empathy, which more closely resembles perspective-taking or mind-reading than feeling into and sharing another person's experience. The insular cortex is activated when emotions are observed in others, including pain and aversive reactions to unpleasant foods. However, while emotional empathy appears to be a felt sense in the body, cognitive empathy can be a "mental calculation" of another's experience that doesn't necessarily involve shared feeling.

Compassion and emotional empathy have been used interchangeably in recent decades, but are technically defined differently. The scientific difference is summarized by researchers at the Max Planck Institute for Human Cognitive and Brain Sciences in Leipzig, Germany, and the Swiss Center for Affective Sciences in Geneva, Switzerland, as follows: "In contrast to empathy, compassion does not mean sharing the suffering of the other: rather, it is characterized by feelings of warmth, concern, and care for the other, as well as a strong motivation to improve the other's well-being." That said, the two can certainly coexist in human experience.

To shed some light on compassion, which is the more action-oriented of the two, I interviewed University of California, San Diego, neuroscientist Karen Dobkins, PhD, who lectures and teaches courses on connecting more mindfully with others. We covered a bunch of topics, including impression management, compassion, and taking responsibility for your role in a conflict.

SB: What are some common barriers to mindful connection?

KD: Recently I gave a talk on impression management. Impression management means that you're trying to manage the impression that people have of you. And humans do this; primates do it too. And if you look at the biological origins of why it is that humans try to impress people, think about a peacock. In other words, on some very basic level, we are like peacocks in that part of what we do is impress so that we find a mate. It's part of the mating game to tell the person that you're interested in: "I'm pretty, I'm healthy, I'm smart, pick me." And also in the sense of belonging, you can go back to the days of cavemen where if you were not a team player, if you didn't have traits of value to your clan, you would be kicked out into the cold and you would die because you couldn't survive on your own. You'd be eaten by a saber-toothed tiger. You'd freeze. So you had to be appealing to your clan.

So I start up my lectures on impression management saying, "Yup, impression management, it's part of our genes, it makes sense. We need to do it." But the problem is, in my opinion, that the stage is getting bigger and bigger. There was a time when humans only had the written pen and then they had newspapers where information could get out and people could learn about other people. Now there's the Internet and now you're onstage all the time. Now, with how big the stage is, the pressure to impress everybody and be something to everybody is so strong. We're getting way too focused on the outward stuff and impressing. And all of that comes at a cost. If you're spending so much of your effort on impressing and on the outside stuff, you're not using that energy to make true connections with people which have nothing to do with your job title or how much money you make or how pretty you are. In the modern age, it's getting harder and harder to connect on a kind of a deep, soulful level because we're putting so much energy into the impression part.

SB: Yeah, that makes sense on a societal level right now. It's harder for everyone to focus internally or authentically because our attention is just constantly pulled outward all the time.

KD: Even if I wanted to look up how to make friends right now. Let's say I want more than this impression management stuff in my life. I want deep connections. What do I do about it? I go on the Internet, I'm going to find like five million sites that each tell me how to make good connections. And I feel overwhelmed because it's too much information. I just want my Aunt Betty to sit me down and explain to me how to make a good connection. So there's just information overload—all good stuff, but even better is to just sit me out in the woods, take me off the grid, let me get back into nature, into my body, into simple living. But there's no schooling for that. Schooling is all about learning how to navigate the Internet and get information. Where's the class on standing up tall, not getting confused by your mind, and really connecting deeply with yourself and others? There's no course on that.

SB: If you're lucky, you sort of receive it from parents or mentors when you're growing up . . .

KD: I've been a big advocate for it at UCSD, where I teach a class on how to be a human. I've been doing it for six years and it's all about this stuff because I just can't believe that children get through high school. We all think it's so important that they take math and science, and some of that they'll never, ever use again. I'm like, here's how you get connected to yourself and here's how you actually have a truthful connection with another person. I can't believe we don't educate people on that.

SB: Would you be willing to share some of some of the tips that you use in your course to help people connect with themselves and others?

KD: So in my course the first half is about your relationship with yourself and the second half is about your relationship with other people.

The first part of the course on your relationship with yourself is really all about how do you deal with your anxiety and depression? How do you manifest what you want? How do you stay on a path that has heart? I always ask students, "Where are you in your element? Are you on a path that has heart or are you on a path that you think you should be on?" We talk about mindfulness and some Buddhist principles, and we also, in the first half of the course, do body scans and meditation.

Most people go, "Oh, I'm in my head," because we really associate sort of with our mind and our thoughts. But the truth is, you've got a whole body below your neck that's got lots of intelligence. And I love to use the word *intelligence* for the rest of the body, because we usually use the word *intelligence* to think of our brain, but oh my goodness, your liver, your heart, your spleen, your lungs—each one of those organs is incredibly intelligent. And when you sort of honor that and realize there's a whole body to be listened to, you can get out of your head. Because we're constantly talking to ourselves . . . everything we think we say is meaningful, and it's like, no, it's just a noisy mess most of the time.

But the second half of the course, which is the best part of the course, has two parts. One is about learning compassion and one is about learning how to take responsibility for your contribution to a conflict.

The compassion part is about not judging people. Not because we're trying to be nice. Compassion is not about being nice, and I always say in my course: this is not a course on learning how to be nice. In fact, learning how to be nice, unfortunately, can turn into a Band-Aid and a kind of a veil for really doing the work because you decide, "Oh, I'm just going to be nice."

But oftentimes being nice means you're not telling the truth, you're not authentic. So I go, this is not about being nice. Compassion is realizing that as soon as you start to judge somebody, whatever adjective you have assigned to that person, you are that thing too. And to deeply understand how ridiculous it is that we separate ourselves. I use the word *separation* a lot, and separation is a type of suffering. We separate from people and we do it in the name of, "She's so different from me. She's so inconsiderate, but I'm not." And you kind of go, "Whoa, whoa, whoa. So you're never inconsiderate?" And what you realize is that everybody's inconsiderate sometimes.

But what humans do is we slice and dice. We go, "Well yeah, I know there's a continuum of inconsideration, but here's the criteria of it and I happen to be conveniently on this side, which is okay, whereas, oh, she's on that side of the criteria, so she's the one who is really inconsiderate," which is just crazy because who made you the authority to decide the criteria that isn't okay?

SB: And compassion helps us get around that.

KD: I get people to deeply look at the fact that we're all the same. You know, we all have negative traits. Some people have more of this negativity. But it's all the same soup. And once you realize that—and this is really important—it's about *connection*, because once you stop judging people and pointing fingers, you instead go, "Oh, I get that. I get what's happening for that person, because I'm like that too." And then instead of the separation you usually have when you point, you connect.

And it doesn't mean that you put up with other people's bad behavior; it doesn't mean that at all. It doesn't mean that you let people walk all over you. You can absolutely have your strong boundaries and say, "I will not accept this behavior," but that's completely separate from the deep understanding and love and compassion that you feel for that other person because you really are just like them.

> For anyone with a complicated relationship to their family, 2020 was a difficult year for connection. In October, not long after we spoke, I saw that Karen had a public lecture on YouTube called "Compassion for the Perpetrator," which she'd released earlier in the year.

In it, she discusses how societies heal after massive human rights violations, and how individuals heal after major and minor traumas. She makes the case that all humans have a dark side, and raises the controversial possibility that recognizing this may lead to healing for both the victim and the perpetrator. Toward the end she quotes Thich Nhat Hanh: "When another person makes you suffer, it is because he suffers deeply within himself, and his suffering is spilling over. He does not need punishment; he needs love. That's the message he is sending."

BODY ONENESS AND MOVEMENT SYNCHRONY

At the University of Oxford's Social Body Lab, Emma Cohen, PhD, Joshua Bamford, DPhil, and Bronwyn Tarr, DPhil, study the effects of dance, music, and movement on social closeness. Their work, which largely focuses on movement synchrony, has shown that behavioral synchrony among rowers is correlated with elevated pain thresholds; that synchrony and exertion during dance (including silent disco) independently raise pain thresholds and encourage social bonding; that movement synchrony forges social bonds across group divides; that interpersonal movement synchrony facilitates pro-social behavior in children's peer-play; and that movement synchrony exclusively guides infants' social choices after 12 months of age. Cohen has even found that synchrony promotes social connection in virtual environments, suggesting that moving together online may be a powerful way to stay connected.

It's no accident that synchronized movement has become such a central part of so many cultures. Serving an adaptive function, activities like dance, sports, and music making encourage group cohesion by triggering the release of endorphins that influence social bonding. "By acting on ancient neurochemical bonding mechanisms, synchrony can act as a primal and direct social bonding agent, and this might explain its recurrence throughout diverse human cultures and contexts (e.g., dance, prayer, marching, music-making)," Cohen says. Whereas other mammals primarily bond only with their close kin, humans have evolved to be capable of bonding with both kin and non-kin. "Dance may have been an important human behavior evolved to encourage social closeness between strangers."

Joshua Bamford's work centers around how music and dance have evolved as human behaviors. Specifically, his research investigates the social bonding effects of synchronized action and tries to pick apart the mechanisms that underpin these effects. He uses a range of methods, from motion capture to eye tracking, and draws upon general principles in cognition and perception. I spoke with him to learn more about movement synchrony in the context of social connection.

SB: Which types of movement has your research shown to boost social connectedness the most?

JB: My own research has focused on the effects of coordinated movement in music and dance contexts. We tend to find that tight, temporal coordination or synchrony generates the greatest feelings of social connection between participants. It also influences their behavior, as people will turn to focus their attention more toward someone who is moving in synchrony with themselves. In these contexts, the music is used to provide a kind of scaffolding for movement and helps people to get in time with each other, but it's probably the experience of synchrony with another person that leads to the social bonding.

SB: Do you have a theory about why movement has such a powerful prosocial effect (biological, evolutionary, philosophical, etc.)?

JB: I suspect that people like to move in synchrony with others because it makes it easier to understand them. I base this on the principles of active inference, which suggest we make predictive models of our environment, and that these underpin our actions. Any uncertainty in our models we resolve either through refining the model or taking action to change the environment so that it fits our model. When we interact with others, we dynamically model each other's actions. Achieving synchrony requires aligning those models of each other, until we can easily predict the other's actions based on our own. This process of predicting each other's actions creates a profound shared experience in which we actually make ourselves like the other, and which can form the basis of understanding the other.

SB: What are some practical ways to incorporate movement into one's daily life to boost connection?

JB: I like to take the advice of Nietzsche: "And let each day be a loss to us on which we did not dance once!" I realize it's harder in our current times of physical distancing, but the great thing about music and dance is that you don't have to be in physical contact with each other. You can dance with someone on the other side of a room, or you can put a YouTube music video on and dance along in your bedroom. You don't need anyone to dance with, as we can still form a kind of parasocial connection with music when we interact with it. Even something as simple as going for a walk with someone allows for an opportunity to synchronize your movement with theirs, as you have to coordinate the timing of your steps to walk side by side. Dancing is really just stylized walking.

Still, I wanted to know what it is about synchronous movement, in particular, that brings us closer. According to Alan Page Fiske, whom I reached out to again, it's more about the synchrony than about the movement.

"Equivalence among bodies makes people socially equivalent," he told me. "The most powerful and pervasive mode of intensifying communal sharing is to give people the feeling that their bodies are one."

With regard to synchronous movement, it appears to be about several bodies becoming one. "It seems to have something to do with the perception of being one body. When my left foot takes a step, everyone else's does too, so our bodies perceptually and kinesthetically merge."

But the social bonding benefits aren't exclusive to movement, Fiske says. "The sense that one's body is equivalent to the bodies of others makes one feel socially equivalent. This sense of equivalence occurs when people feel the sameness of their essential substances, body motions, or surfaces."

That means we can also feel close to others when we perceive that they share the same blood or genes; consume food and drink together; share tobacco or other drugs; mix or shed blood together; kiss and hug each other; exchange body fluids during sex; get tattoos together; wear the same clothing; have the same skin color or hair type; or share a disease, disability, addiction, or trauma.

"One body is one social self. People connect to each other by assimilating their bodies to each other, making themselves one substance together."

Body oneness also leads to more "peak" bonding experiences because the feeling of closeness arrives more suddenly.

"It is the sudden intensification of communal sharing that evokes kama muta [peak experiences of closeness]. The steady, taken-for-granted sense of being one body, one being, does not activate it—the steep increase does." More than any other kind of action, body oneness causes a steep increase in communal sharing, evoking kama muta more powerfully and more consistently.

Having a lot in common with someone doesn't necessarily create a strong bond; having a shared bodily experience does.

That said, Fiske concedes that not everyone benefits from body oneness in the same way. "All kinds of factors may distance people from it, block its effects, or completely reverse them." To have peak bonding experiences from movement synchrony or body oneness, he says, "people have to be prepared for and disposed to relating communally."

> When I lived in Portland, my boyfriend and I would watch *The Wire* together in the basement of our apartment, where our landlady had set up a projector. We had a little synchronized dance we did with our heads to the intro song by Tom Waits, nodding them up and down to the beat, first to the left, then to the right. We took the dance with us to Australia, where we were always on the move, camping for a month in Tasmania and continuing to watch *The Wire* on his iPad in our tent at night with a couple of warm beers. Movement is how we first connected: meeting at a college party in Ohio, dancing together for hours, still there at 5 a.m. after everyone had left. After graduation six months later, he moved to Oregon so we could be together. We ran together, too, moving our bodies in synchrony up the slopes of Forest Park in northwest Portland, the wan sun on our backs. In warmer weather, we would take books to a bench by the Willamette River to read. That's one of my favorite memories, reading next to each other in the middle of our life together. Even in memory, these feelings of connection are stronger when I think of moments of movement and oneness between us. Reflecting on the present, it seemed that the same went for memories of dancing in Berlin's clubs when they were shuttered during a pandemic, or lifting off and touching down in a new country when traveling was easier. Our bodies can be connected even in our thoughts.

MINDFULNESS

Jonathan Gibson, PhD, is a professor of psychology at the South Dakota School of Mines and Technology. His work looks at the intersection between interoception, mindfulness, and meditation. His article "Mindfulness, Interoception, and the Body: A Contemporary Perspective" was published in *Frontiers in Psychology* in 2019 and is one of the most insightful and thorough explorations of these topics I've read. The article acknowledges the complexity of defining and adopting mindfulness techniques and advocates for "attentional style" as a more useful and constructive concept. In it, he writes:

> In a healthy population, simply attending to the body has shown to promote a number of benefits. For others, attending to the body can elicit discomforting sensations to where the participant cannot feel comfortable in his or her body. If we grant the body a central role in this process, then attentional style can function like a focal point from which to investigate complex interoceptive signals. Each meditation with its given attentional style/focal point can reveal different perspectives or dimensions of our embodied nature.

I spoke with him to hear his story and learn more about the distinctions between mindfulness, interoception, and meditation. As a prelude to this interview, I would like to acknowledge the non-Western origins of both mindfulness and meditation practices, and clarify that Gibson approaches these subjects not as an expert in contemplative practice, but as a scientist and researcher applying a neuroscientifically informed interoceptive lens to these techniques.

SB: I'm interested in your interest in interoception. How did you get into this area?

JG: That's a good question. Probably similar to you. When I finished my undergraduate degree, I was really disillusioned with psychology. I'm more of a big picture kind of guy. I wanted to see the forest and I never saw the forest; I just kept seeing trees. And so when I got in my master's program, I answered a lot of those methodological and theoretical and philosophical concerns. But as soon as I satisfied those, I had this kind of inchoate sense pulling me toward the body. At first I was like, I'm interested in Buddhist psychology. And then I realized that wasn't it. And then I started looking at the psychophysiological research and started reading Stephen Porges's work on the polyvagal nerve. And the question from which my interest emerged was from looking at these wisdom traditions and how they supported the notion that the body is a source of knowing, source of wisdom. And so I go, well, if that is the case, how do we get access to it and what happens when we do? And then I stumbled upon the interoception literature. In my PhD program, we were given a lot of latitude to explore our own interests. My dissertation advisor had no expertise or even understanding of the interoception literature. So I felt like I was on the frontier kind of trailblazing.

Again, I was kind of disappointed. Prior to Mehling's piece in 2013—he was the first to argue that interoception is a multidimensional construct; it's not this uniform construct that we can measure with a singular modality, whether it's the heartbeat detection test or the water load-bearing test—prior to that, the literature was kind of stuck in like, I don't know if "purgatory" is a good word for it, but from my vantage point, it was in a methodological straitjacket. [Researchers] were consumed with going, "We have to be able to measure this." And so they were stuck in viewing interoception in one particular way, which was really limiting. This is a very nuanced capacity. You and I can both focus in on our chests and ask, "Can I feel my heartbeat at rest?" Well, what does that tell you? I mean, it's going to be different for every individual, right? Being able to detect your heartbeat at rest doesn't tell us a whole lot. So after I finished the literature review, I started working on collecting my data and I had a pilot study and then I had two different phases. It took about six months. And most of my data was qualitative, which gave me a lot of the insights that led to my eventual paper I published a few years ago.

To answer your question, something was pulling me—I even acknowledged that in my dissertation—it wasn't like I could pinpoint or define it, but something was pulling me toward this direction, looking at how we can reconceptualize interoception and how we can benefit from it. Because from my own personal experience, I would do this all the time—I would check in and feel how my body would respond to certain things. One of interpersonal neurobiologist Dan Siegel's techniques is that if you say the term "yes" and just pay attention to how your chest feels, it kind of opens up, and with "no" it will feel constricted. And that's something I've been doing for half my life. I would just kind of check in with that.

SB: The cool thing about it is that when you describe what interoception is to people, they get it. They don't need much of an explanation. We all have this experience of using our body as a resource in this way.

JG: Most of us do. I teach on a STEM campus. So it's almost primarily engineers and they're very left brain dominant and a lot of my students are really detached from their bodies.

SB: How can you tell? What does that mean?

JG: Well, I think, generally speaking, most people in my culture are kind of detached from their bodies. They view their body as a machine that carries around their brain. I've been asking my students questions like this for a long time: I'll go, "How do you feel about that?" And rarely do they answer that question correctly. They'll tell me what they thought. And I'll say, "I didn't ask that question. I said, 'How did you feel about that?'"

So let me give an example. When I teach my Intro to Psych course, we talk about mirror neurons and I show a video. And in the video they're describing how mirror neurons work and how they play an important role in empathy and feeling into other people's experiences. There's a guy in an MRI machine and they show him pictures of people making faces, smiling or frowning, angry or disgusted. And I let them watch that for about 10 or so seconds. And then I stop and I go, "Did any of you notice how you felt when you saw those faces in the image?" And nobody pays attention to that. And so I go, "Okay, go back and let's pay attention to how you feel when you see those faces." And then about a third of them go, "Yeah, I felt something," but two-thirds go, "I don't feel anything."

So we have to pay attention to that, because there are signals being sent from the body to your brain that do inform and constitute your psychological state. It just doesn't cross the threshold of awareness. Some of them are naturally attuned to that. But for a whole bunch of my students on this campus, they are not naturally attuned. It takes effort to start paying attention to those feelings and sensations.

SB: I think you'd be interested in the work of Kristina Oldroyd. She's a psychologist at Utah Valley University and studies the developmental origins of this. She's interested in, you know, do some of us naturally have this ability to sense these internal processes or is it a function of how we're raised? But, of course, culture plays a big part as well. Especially for boys in America, it seems there's this pressure to not feel at all or not admit that you feel. But it does sound like it can be developed.

JG: Yeah, I think so. There may be a set range for some people, some people can have a little bit more accurate interoceptive awareness than others, but you can always improve.

SB: Do you think it should be part of educational curricula? How would you teach it?

JG: I think it's almost as valuable, in some ways more valuable, than any traditional curriculum. It's right up there with reading, writing, and arithmetic. Because learning how to attend to your body, to manage your emotions, understand yourself, and understand others is something that's so profoundly deficient in our culture. I don't want to extend myself too much, but it could have a huge transformation on so many people in society if they were taught to pay attention to their body from the third grade through college. Just five minutes a day, learning how to breathe.

I do teach it. Like the example earlier with the mirror neurons, I ask how they are feeling. And I meditate in all my courses. So just in my intro course, we do a four- or five-minute breathing exercise. I stress them out first. I show them a YouTube video—and it's just an audio recording of an infant crying—and they only need 30 seconds of that and their stress levels are really high. So I'm like, "Okay, now we're going to breathe and you're going to feel the effects of how just breathing increases vagal tone and dampens sympathetic or fight-or-flight response." It increases their parasympathetic activity so essentially those stress signals never get into their brain. It allows the brain to kind of recalibrate, and they can feel the effects in just four or five minutes. But then I have other courses where I explicitly teach them how to meditate and they have to meditate daily for six weeks.

I have a pet course that they allow me to teach here. It's my mind and body course. [Students] have to do six weeks of meditation and I let them choose which meditations they want because I'm still kind of curious whether there are differences between different meditations. They have to learn all about it, and I have them summarize the experience at the end. I have two or three other courses where I make them meditate because it's just so effective at managing the stress, and they need it. This is a pretty rigorous campus and the curriculum is pretty rigorous for them.

I read your blog about social interactions [and interoception], which was really fascinating. I hadn't looked at that yet, so that was new to me, but it makes a lot of sense. So many of these students are kind of shy and introverted and they don't have that social connection that others do that they really need. And if they don't have that connection to help them manage their stress, they need something else. Unfortunately, so many students just drink instead. That's how they deal with their stress. So I'm trying to provide them with a different set of tools that they can use to manage these challenges.

SB: The distinctions that you drew in your paper between different attentional styles were really interesting to me as well, especially the difference between the focused attention and open monitoring and going back to this point of not necessarily wanting to pay more attention to body signals, for example in the case of being a traumatized person. You might just end up triggering a conditioned response unless you're using this nonjudgmental, safer, open kind of monitoring style. There was also one point you made about how sustained attention is important in particular, and I think it is just one line in that section, but I found that to be a really good point. Because you can talk about paying more attention to what's going on in your body as a good jumping-off point for better self awareness, but it's not just a one-and-done matter of noticing the signal and understanding suddenly what that means and contextualizing it in your personal history. It takes longer than that, and it sounds like sustained attention is really important, but also not possible for everyone unless you have this kind of remove or disidentification from it. So I'm interested to hear more about that—the open monitoring style—because that seems especially useful in the trickier situations.

JG: I was heartened to see that the literature supported that because when I originally came across this, it was in my own research. Like I said, I did a pilot study and then I did two phases. One was more of a comparison, and then the second one was looking more closely at people's experiences and looking at the differences between a focused breathing meditation and a body scan meditation. A bunch of my participants would go, "Yep, I'm doing the body scan; there's my feet, nothing there." It was kind of surface level. They never got deeper. What I did see is that even though they never got deeper and they were just doing the breathing, they still experienced the same effects that we see in the mindfulness literature with attention regulation, being calmer, controlling their emotions better. There was a change in perspective of the self. But what some of my students showed me is that after that surface level, which may be the foundation I was mentioning earlier about interoception, there is a deeper level that you can gain access to that can reveal a lot more about yourself and that requires a different attentional style. And that's kind of what I was trying to argue in the paper: Yes, you can just focus on your body and do a body scan, not mindfully, and that will get you so deep. Then mindfulness might get you deeper. And then focusing can get you even deeper. It just depends on how you view your body as that kind of focal point. There's no way this is going to come up if you're just breathing without that deeper dimension. You have to alter your attentional style in order for that to open up. But to your point, it has to be sustained. It takes time for that to open up.

SB: And one potentially more efficient way of doing that is taking MDMA, perhaps. The open monitoring attentional style, in general, is very similar to descriptions of clinical studies where MDMA allows for people to revisit their traumas because it's from a safe distance. It's an entactogen as well. You could argue that the same thing could happen on classic psychedelics, but MDMA in particular seems to be most effective for that kind of treatment. And I wonder . . . it must be engaging the interoceptive network. That has to be involved.

JG: One of my gripes is oftentimes they'll talk about the interoceptive network, but they'll just call it the "center for emotional processing." So it's like they're dumbing it down for the layperson when we really need to articulate and argue and bring into awareness what the insular cortex is really doing. My kid bought a little book about the brain and I can grab it if you're interested [grabs book]. And this is the only book that actually talks about the insular cortex right here. This blew my mind. I teach seven different courses on campus. None of my texts, at least the texts you can use for development or abnormal [psychology] or anything like that, none of them talk about the insular cortex. And it's like, this actually is maybe just as important as the prefrontal regions. And I really wish I kept the study, but it was a study published right after mine and they made a big deal about the open monitoring attentional style not being driven by the prefrontal regions but by the insula. So that's why I think that the core of mindfulness and all these other meditative techniques is really just interoception.

SB: Well, I guess it is difficult to study, being deeply hidden within the brain. I think the Allen Institute in Seattle last year did one of the first single-cell analyses from von Economo neurons in the insular cortex. But you're right. And if you're considering anything in terms of the salience network, it's going to come up. Increasingly, the psychedelic literature is talking more about the salience network and less about default mode network and ego dissolution, and my personal opinion is that it's much more about social relationships than anything else. It's about the bodily self, but the bodily self can't be separated from the social self. We're not isolated units in the world. We're constantly interacting with other people. So you have to also look at the insula's role in social cognition and how it interfaces with interoception as well.

JG: Well, that's the thing. And so I hope that the interoception literature doesn't get pigeonholed like mindfulness did. Because the insular cortex and the network that includes the cingulate and somatosensory cortex—it's processing more than just awareness of internal bodily sensations. It acts as a switch between the default mode network and the executive control, and that doesn't fit into the interoception literature. That's something different. So I hope the interoception literature can provide a foundation, but I hope it goes beyond that.

One of my qualms that I brought up to Laura Schmalzl and Wolf Mehling at a conference a few years ago was why haven't we included any of the body and its functions and peripheral systems into the interoception or contemplative literature? We acknowledge that the body sends complex signals, but the brain is representing those signals, and that's where we're stuck. But there's a good mountain of evidence that the peripheral systems actually change, which change the signals being sent to the brain. So it's not just how the brain is representing those signals, but there's actual changes in it. And if we don't include the body with the interoceptive network, then we're only getting part of the equation. So I would hope that a deeper foundation would include the enteric nervous system, which is the gut–brain axis, or the vagus nerve or maybe even the heart as it seems to play a role in our psychological functioning state. So I hope we can extend beyond the interoception literature and include the body as an important piece, maybe the foundational piece, from which we investigate all of these topics.

SB: Have you heard that people who have better interoception have a better ability to draw a line between themselves and others in a visceral kind of way? It's this idea that once you are more in tune with your internal state, and you have a solid representation of your bodily self, it's easier to—maybe in a situation where you are being required to express empathy for another person—it's actually an advantage to have this self-other distinction because otherwise you might just get pulled into the other person's emotions and not be able to distinguish yours from theirs. It also relates to trauma survivors in the context of situations where they can't separate their sense of self from whatever is being thrown at them.

JG: There is an article that was just recently published talking about how classical conditioning shapes memory and how so many people are just responding to a conditioned memory and not to what their bodies are actually experiencing. And then on the other side of the coin, what you were mentioning is that we share experiences. That's what empathy is. To the researchers' credit, they mentioned the insula as having been identified as the hub of empathy 15, 20 years ago. And they recognize that people who are really empathic have an active insula. They can literally feel into someone else's experiences and they kind of represent that in their own body.

But to your question of how important is it to differentiate your feelings from someone else's feelings? [This affects] everyday interactions. You can have a discussion with your partner, and if they start to get heated and then you start to get heated, you feel the same thing that they are feeling and you fail to recognize that you're coming together in an objective moment where you're both representing each other's experiences.

SB: That's a good way to put it. So in that sense, interoceptive training could probably help with that. But then there's also these feelings of connectedness and unity in the psychedelic discourse, potentially driven in part by the insula, which is a really interesting paradox. Because if it's true that the interoceptive network is engaged during these positive psychedelic experiences, it seems at odds with the idea of the self-other distinction being so important.

JG: Well, if you look at interoceptive awareness, it makes sense to me. If you look at these cultures in these traditions that have practiced meditation for a long time, part of their deeper assumptions about how the world works is interconnection. We're all part of something greater. And when I talk to my students, and we define the self, I go, "Okay, how do you define the self?" It's usually skin encapsulated. But in Japan, the definition of the self is one's portion of the shared life space. They recognize that their selfhood is connected to something else.

And that's interesting because I encourage my students, when it's warm enough out here, to go meditate in the Black Hills. Go to a little mountain range, just go meditate out there. And it does kind of induce—more easily or readily than in other environments—this feeling of sentience, that everything is alive. And that we're all part of this, part of something greater. I've had students describe that, and it's very similar to what we see in psychedelic experiences. It makes sense to me that we're activating the same kind of neural circuits involved in those processes and those perceptions.

SB: Right. Well, and maybe there's even a step in between, where you actually have to feel a stronger sense of self before you feel connected.

JG: Yes. That was the point I wanted to make earlier. You have to have a stronger sense of self before you can really differentiate between self and other. And that's one of the first pieces that comes together when you increase your interoceptive awareness. I had a lot of students say they felt like they found themselves, knew who they were, even after five weeks.

SB: Would you be willing to talk about how you've used interoception in your own life at all? I mean, I can think of many examples in my own life, over the past year even, where it's helped immensely just with emotion regulation and just knowing what I'm feeling and what to do about it.

JG: I've kind of always lived my life this way. I never make a big decision until I consult what I would call my "felt sense." There's a way of knowing, "Is this the right thing for me to do?" And I would just kind of wait. And if it wasn't clear for me always, I would wait until it did become clear and it just kind of felt right. It would resonate.

I have four children and I work with them a lot. And we develop and cultivate interoceptive awareness. My oldest has a tendency toward feeling like he has to win. Maybe there's some insecurities there. He's the oldest and he has to win or his value is diminished. And he has caused some strife with his siblings. He was kind of losing control and being bratty and mean and causing lots of fights. And what happened was I tried to get him to use that focusing technique and I go, "Okay, just pay attention to your body." This was after the storm kind of settled because he was still angry and I had him breathe and calm down. And then I go, "What's going on? Why are you feeling this way?" And what happened is he just kind of opened his eyes and he goes, "It looks like there's a cloud in my chest and there's this dark storm." I go, "So just ask it what it needs. What's the next step in the healing?" He just sat there and it took a while—and to his credit, he's actually really good at meditating for a young kid—but there was a noticeable, palpable change in him and he goes, "There's light in my chest now instead of this dark spot." So that's what I was talking about earlier: that deeper dimension that you can't get to if you don't start asking and reflecting in that open monitoring style of attentional practice. But it just opened up for him and his behavior changed and it was different for months. He was no longer trying to pick fights with his siblings.

So that's how I use that deeper bodily knowing in a kind of direction that orients our lives. We never make a big decision until we ask, "Is this the right thing for us to do?" And we just kind of wait until we feel it. It's hard to describe beyond that. It feels open. It feels right. Eugene Gendlin says there's no way you could ever know that something feels wrong unless you knew what felt right. When Carl Rogers was talking about organismic valuing, he was talking about the same process. There's an embodied way of knowing. It knows the direction of how to actualize things; it knows what you need in your life. You have to follow that. And so the way I view these different meditation techniques is to tap into that deeper bodily way of knowing. And there are different attentional styles that you can use to do so.

SB: That's also very empowering. I mean, you're giving that power back to the person. Even as a parenting tactic, that's great. You know, you're not telling your kids, "Stop fighting, stop doing that." You're really handing the responsibility to them.

JG: Most kids don't have this. And that's why I think it would change their lives so much, because it would give them the power to have control over their own lives.

> A paradox was starting to surface from my interviews, something that seemed to run counter to mystical narratives of ego dissolution and anthropological theories of oneness: merging is not connection; merging is the loss of agency and awareness necessary for the experience of connection. To connect, deeply and fully, you need a stronger sense of self, and you need that self to be recognized—not subsumed—by another.
>
> In my drinking days, I might fly out of my body and into someone else's head during a sober conversation, but alcohol seemed to ground me in the feeling of my body, which helped ground me in my mind.

The way it channeled through my limbs, and made me aware of my inner experience, translated to psychological self-validation. If I was in an interaction with someone, having some kind of pleasant bodily experience, that sensory reality became something I was certain separated me from the other person. There was no way to be wrong about what I felt, and no way for anyone to tell me I was wrong. Free of the pressure to merge my experience with theirs, I could connect from a more authentic place.

Over the past few years, I've become more familiar with my own internal landscape, as it is. I now think of it as a living, breathing entity of its own, separate from my mind. It has its own substance. It speaks its own language. I listen. Even as I type this sentence, I feel a calm hum rising in my chest that tells me, Me, here, I'm right here. The more I pay attention, the louder its voice becomes.

TRUST

Kelsey Blackwell, MS, is a cultural somatics practitioner and author of *Decolonizing the Body: Healing, Body-Centered Practices for Women of Color to Reclaim Confidence, Dignity, and Self Worth* (New Harbinger, 2023). As a facilitator, coach, and speaker, she has brought abolitionist embodied practices to such diverse groups as riders on Bay Area Rapid Transit trains, students at Stanford University, and the offices of LinkedIn. She works one-on-one with clients as well as leading the eight-week Decolonizing the Body group program. Blackwell believes working toward personal and collective liberation must be impactful but also bring joy. She lives in San Francisco.

I spoke to Blackwell about a range of topics, including body trust, which is not only a folk concept but a scientific construct, used to measure interoceptive awareness. More than any other facet of interoception, low body trust has been correlated with depression, suicidal ideation, eating disorders, loneliness, and other ailments.

SB: What does body trust mean to you? How do you define it, practice it, teach it?

KB: For me, body trust means honoring my embodied experience, which seems fairly simple, but we aren't actually taught how to be embodied or how to honor what our bodies are expressing or feeling or communicating. And so much of what we see modeled and then start to do ourselves is disconnect from our bodies so that we can do the things on our to-do list. Our bodies become locations of annoyance or distraction, these things that we can't quite figure out, that feel a little messy, that we try to push down or push away what they're presenting because we aren't really taught how to work with it. So for me, trusting the body is actually trusting that my embodied experience is valid and that it's worth listening to.

More than that, it is understanding that the body connects or communicates through the language of sensation. And we have a choice in how we relate to those sensations. We can tune in or we can push them away. We can sort of subconsciously feel sensation and then go into an embodied response. But the approach of trusting the body means that I'm bringing those sensations into my awareness. And there's some sense that whatever my body is presenting, whatever those sensations are that are arising, that they ultimately are in service to my life, which is, I think, somewhat radical for folks to consider because oftentimes it feels like that is not the case. The sensation is getting in the way of things. And so trust in the body is the opposite approach. These sensations are here for a reason. And when I tune into them and learn how to work with them, ultimately they're going to bring me toward what is life-supporting or life-affirming. When I say life-affirming, I mean that I get to feel that I am safe, connected to myself, connected to others, and that I have dignity. Those are the three things that all bodies require: that they are safe, connected, and have dignity. Dignity means that I deserve respect and that I don't have to do anything or prove myself in order to have that. The sensations in my body are indicating to me how to have those three things happening in my life.

The other thing that I think is really important for all bodies, regardless of how they're coming in to work with me, is that building this relationship with the body where we can start to feel for how our resilience lives inside of us. So the body is not just a location of aches and pains and panic and anxiety, but that it's also in the body we can start to feel the places that feel strong, that feel grounded, that feel resourced, that feel available. And we oftentimes don't pay attention to those places because they feel "fine" or "blank" or our mind doesn't even linger there. But when we start to build a relationship with the places that aren't paining us, we are also starting to recognize the body as not just a location of problems, but also a location where we can access ground and safety and strength.

SB: How can we think differently about discomfort, for example the moment when we want to honor our feeling (and therefore ourselves) but hesitate because of conditioning or anticipated rejection from others and/ or the world?

KB: I love this question because it describes kind of a common situation that a lot of my clients will experience, which will bring them into practice, which is that there is a way of showing up in the world that aligns with their values that they feel that their body isn't allowing for. So it might be like, "I'm in this meeting and I know I have something to say and I know I have an idea that's worth sharing, but I find that I keep holding back or I can't quite get myself to say the thing that I want to say." And then there's this internal feeling: we can sort of self flagellate, like, "Why didn't I say the thing? I told myself that this was going to be the time," or "Why didn't I stand up for myself when that person said that thing?" We start to feel this inner tension within ourselves where it can feel like the body is actually holding us back. The mind says, "I know who I am and I know what I stand for and I know how I want to be seen in the world," and yet in the moment I can't quite get there. So that tension can sometimes make us battle ourselves, where there can be some negative self-talk or we can put more pressure on ourselves to do it better next time.

Some of this negative talk is really connected to our conditioning. But the first thing to do when we are feeling that moment where there's something in our body, where we're feeling that moment of discomfort or that moment of, "Oh, I want to take a risk here," or whatever it is, is to notice. And this is part of trusting the body and honoring the body's wisdom—first noticing, "Oh, I'm having sensation in my body, and this is true. I'm not going to push this away. I'm not going to try to push past this to do the thing. It is part of what is happening in the environment right now." And then the more we're able to notice sensation, the more we start to see what our patterns are.

I sort of think about it this way: the body is like the night sky, and sensations are stars. We have all these different sensations. If you look up at the night sky for the first time, it's kind of overwhelming and awe-inspiring, right? But if you look at it over and over again, you start to see constellations. You start to see how you can recognize maybe a planet. You start to see how it's in a different place than it used to be, or how the moon is in a different phase, right? It's the same with the body. The first time we start to feel sensation, it's like, wow—it's like looking at the night sky. Oh, my gosh, right? That's the first piece of the work, to just be with that immense expanse of information.

As we start to feel for sensation more frequently, we make those constellations, which is when we start to see our patterns. And seeing our patterns means that we also start to understand where those patterns come from. I see why this embodied response is showing up. This embodied response is afraid that if I say something here that I'm going to lose a friendship or I'm going to be ostracized by my colleague, or I'm not going to be liked anymore, or whatever it is. There are so many different ways that this can show up depending on our history, our unique histories, but it's to understand that that sensation is showing up because it's protecting something. So then, rather than saying, "Oh, I don't want to feel this, I'm going to push it away. I'm going to just charge through and do what I want to do," it's bringing in the wisdom of what that sensation is showing you, but letting it know—and this is where feeling for the places in us that are resilient comes in—that while that is part of us, it's not the only truth. We feel that part that feels scared, anxious, angry, that feels whatever it feels—we let that be part of our reality. "Yeah, this is true." And at the same time, I'm also feeling my legs on the ground. I can feel my breath and my belly. Oh, I can feel my hands. I can feel my jaw tightening, and I can let that go. And then we respond from that place. It's holding our complexity, not narrowing to one way of being or another way of being, but instead recognizing that we are like that night sky. As I connect with that larger complexity of what's true in the moment, responding from that place is an act of authenticity.

SB: How do you see social justice informing body trust, or somatic practice in general?

KB: I don't know how you do somatics without social justice. Somatics to me is the practice of understanding that life moves toward life, which is when I was saying that these sensations, even though they're uncomfortable, I trust that they're moving me toward what actually supports my thriving. It's the same idea with the entire practice of somatics, which is life moves toward what is inherently life-affirming. As we are exploring where life is moving us toward, we are inevitably going to encounter what is getting in the way of that. Systemic oppression is not life moving toward life; it is the opposite of that. In my way of thinking, there's no way you can practice somatics without at some point encountering how systems of oppression have shaped your own body, your community, your family. We default toward, "I have this trauma. There's something wrong with me. Let me figure out how to heal this and fix this. Oh, I fucked up again. Damn it. I just need to be better." And it's raising our gaze to see that there are all these systems that we live inside, that our families live inside, that our ancestors have lived inside, that have literally shaped how we are in our bodies over time. I think it takes some of the pressure off in some ways to say, "I see that, yes, I have this pattern. But this pattern isn't just mine. This is a pattern that has been passed down to me. This is a pattern that I've seen reinforced in my community. This is a pattern that I've seen my grandparents reenacting." And it's not so much about blaming, because we're all surviving inside of these systems, but it's understanding where that comes from. When we understand where it comes from, then we're more equipped to relate to it in a different way. We have access to more compassion, for ourselves and for everyone who's surviving in this time.

SB: Would you be willing to give a brief summary of the philosophy behind your book and the course you teach?

KB: Decolonizing the body means learning to trust the liberatory wisdom that lives within all of us. So it's that idea that within all humans, there is this seed of life—that we can, with intention, water and care for—that is really in service to supporting us, to navigate our lives with more authenticity and take risks, to bring forward our gifts, our talents, our dreams. Inevitably, as we feel for that liberatory wisdom that lives inside of us, we confront the ways we're shaped by systemic oppression. So the journey is both celebrating or affirming our deep knowing and healing the ways we've internalized systemic oppression: that sort of internal track that can feel like it can get in the way of our ability to express ourselves or take space for ourselves in the ways that we want to. So one of the things that I think is really essential in this work that I highlight in the book and the program is that as we are affirming our inner knowing, we are inevitably going to be confronted by external or internal voices that challenge or make us doubt navigating ourselves in such a way. But those internal voices are connected to these systems that we live in. And until those systems are dismantled, we're going to be confronting those voices. I'm not going to do this healing work and then suddenly be totally liberated, on my own. That's not real liberation, right? Our liberation is interconnected.

So how do we do this work knowing that we're still living inside systems of disease? It requires that we allow ourselves to be oriented by something larger than these systems themselves. I think about racialized capitalism and white supremacy and the patriarchy. I think of these as all fairly new systems in relationship to a much older system, which I say is ritual and spiritual intelligence and earth; the body really connects us with the earth. That kind of more organic, cyclical, always emerging wisdom that we are inherently connected to. My clients really have a connection to spirit or their ancestors. So part of doing this work means allowing ourselves to develop rituals that can hold us. Something vaster than these systems created by white men. That is something that I really bring into the book and bring into the program. And it is inherently somatic and embodied because we're engaging with another kind of knowing that's definitely more right-brained, creative, starting to trust the unseen. That is the balm that is an essential component to this work.

SB: It's interesting how there's this body of research literature, over here, called interoception, and then somatics I feel like people interpret as more of a practice. In some ways we're using different terms to describe the same thing. How would you characterize the difference?

KB: One thing to keep in mind about somatics is that somatics comes from the Greek word *soma*. And the soma includes not just the body, but also our spiritual intelligence, our emotional intelligence, our understanding of interconnection. It also includes the thinking self, but the soma is our aliveness. What does it mean to be alive? I think about it sort of like the concentric circles of being embodied. As I'm moving about my world, yes, I'm in this body, but I also have this thinking self that's like a little smaller circle and it's in here. And then I have this energetic or spiritual self that's kind of a little bit further out and then the sense of interconnection that I can play with coming in or out. The practice of somatics is what supports the thriving of the entire organism. I think sometimes *somatics* gets used interchangeably with *embodiment* but there is a distinction there that I find really helpful in my work in general, because it allows me to include the spiritual experience.

SB: That's a really important distinction. And the interoception literature is largely focused on clinical populations and awareness of bodily signals or lack thereof. To move into this aliveness territory that you're talking about really reframes it as wholeness of being.

KB: Wholeness of being, yes, that's exactly it. It makes sense to me, the science. I feel like science really likes things that are tangible or that can be proven, and so it makes sense to me that they would be focusing on interoception. Even that might feel a little bit "woo woo." But to include the spiritual component of somatics—maybe that's where science is going. I look forward to that.

> The guitar. I still had C's guitar. About to travel for a few weeks that June, he wrote me, wanting to stop by and collect it. We'd been playing songs and smoking joints in our favorite tree the day before, after talking things through, and he'd brushed my cheek with his lips. The way you'd brush aside a seedling, too small to see. Later, I'd called him and said we could still be friends. We could not, however, be affectionate the way we were before.
>
> I wrote him that I was home but would be on my way to cat sit for a friend in an hour. Once I reached her place, I stood at her enormous picture window, which spanned the length of her flat, watching lightning crack through the sky.
>
> My phone lit up:
>
> *We left already. Thanks for being there though.*
>
> Then, unexpectedly, a second time:
>
> *I've been wanting to say that even if I'm unsure of my feelings and what to think, know, and feel about love, I have had many moments in which I have felt love with you and for you. I thank you very much for your presence in my life.*

MUSIC

Pianist and composer Nicolas Namoradze, PhD, was born in Tbilisi, Georgia, in 1992 and grew up in Budapest. After completing undergraduate work in Budapest, Vienna, and Florence, he moved to New York to earn his master's at Juilliard and his doctorate at the City University of New York. For several years Namoradze taught chamber music, composition, and music history at Queens College. He now pursues postgraduate studies in neuropsychology at the Institute of Psychiatry, Psychology, and Neuroscience at King's College, London, where his research interests include the effects of mental practice and mindfulness on musical performance.

Namoradze, who also lives in Berlin, spoke with me about interoception, mindfulness, and connectedness in the context of musical performance.

SB: How did your journey into mindfulness and musical performance begin?

NN: I was very active on the competition and concert circuit in my late teens and early twenties, but then I decided before graduating from Juilliard that I wanted to take some time off to develop my artistic profile in my voice, find the repertoire I wanted to perform, and compose. And I was also starting my doctorate and teaching at Queens College.

A couple of years went by and that was the time I got interested in meditation, and I started to pursue it quite intensively. After a couple of years in retreat, so to speak, I said, "Okay, I think I'm ready to go back and do a big competition. And if it works out, then, you know, pursue an international career that would result from success on the competition circuit." Because that's usually how it works: that your first big competition, if you're fortunate enough to win it, gets you concerts and you build a career after that. So I hadn't done a competition in many years. The general wisdom is that you need a lot of experience doing them in order to do well in one. And the first competition I decided to do was in fact the largest kind of prize there is. I think it still is the largest competition prize in classical music. It's called Honens, based in Calgary.

SB: Not ambitious at all . . .

NN: Right. And the thing is, it's an especially stressful competition because it's a winner-takes-all system. There is no second or third prize. It's just, you know, the one thing. So I thought, "Okay, I can sit and practice all day." But I decided that I might want to take a more holistic approach in my preparation for something like this, especially because I hadn't done many competitions and hadn't been on the circuit for so long that I thought, "Okay, let me try to do this differently."

So I studied some sports psychology. I did a lot of reading about neuroplasticity. And I leveraged whatever understanding I had of the way the mind works through my study of meditation to really develop mental practice techniques to prepare for something as stressful as this. Not only is it stressful, but one has to perform at the highest level in such stressful conditions that I realized, you know, this is just putting oneself in such an unusual mind state. And so this preparation process was both for how I would prepare for the competition itself, but also how I would optimize my practicing and performance.

I ended up being fortunate enough to win this competition. And I mean, of course, I'm not a large enough sample size to definitively say whether or not I would have won without all of this, but I definitely feel that it's played a very important role and it has continued to play an important role in the very busy and difficult international travel and concert career that I have pursued since, of course, before the pandemic.

SB: What has been the most helpful aspect of mindfulness or meditation for you in all of this?

NN: I really feel that it's been crucial in situations such as, for example, if I have a rehearsal with an orchestra and by the next rehearsal or by the time of the concert, which is the next day, I need to make many adjustments according to how I relate with the interpretation of the conductor or some situation, and I would be able to make advances or changes in my approach much faster using mental practice that I simply wouldn't be able to had I had a couple of hours just at the instrument to practice. So it's allowed me to do things that I think I wouldn't be able to do otherwise, or do them much faster.

But also Solms, and I think Domanski also, talks about this idea that consciousness isn't a product of the brain, but rather the experience of being a brain. It's just a different perspective. We're talking about the third-person perspective and the first-person perspective. And the really interesting thing about meditation here is that it is really the only rigorous way of examining the first-person experience. I mean, that is essentially what it is. Whatever techniques you might talk about, it is really the examination of what that first-person experience is really like and removing misconceptions about the nature of the way thought relates to consciousness. So that is an interesting complement.

I think what is interesting about invoking meditation in all of this—in the question of mental practice and the performing arts—is, you know, how do we cultivate that sense of mindfulness and awareness when we are not able to stop? Because a frequent idea you come across is to just take a moment to stop, to take stock, to just check in with awareness. But when you're in the middle of playing a piece with an orchestra, you can't just stop. The question is, how do you combine that sense of embodied awareness in the midst of this incredibly taxing multitasking that you're inevitably doing in something like this and make sure that concentration doesn't go off the rails when you take a moment to realize, "Oh, okay, so this is what's happening"? And that's a really difficult kind of tightrope to walk. So I've been interested in developing techniques that address that specific problem and finding ways to make the sense of being "in the zone" less ephemeral. Because it inevitably is—it's this wonderful ideal that we catch sometimes and then it disappears. You wake up too much and it goes. So that's what I'm trying to get a handle on with all of this.

SB: Can you describe a little more concretely—if you're in the moment during a performance, applying this technique, what that experience is like for you?

NN: Part of the reason why we practice as much as we do is so that the majority of what we do becomes automatic. You don't think about pressing every single note, you don't think about every single moment, because you've practiced it. This is happening on an automatic level, so obviously there is some mental real estate where you can let your fingers continue doing their thing and you can be juggling a couple of things at the same time.

But it can happen with the piano as it can happen with anything else—you can get lost in thought. You don't realize you're thinking. And then there's a moment when you realize, "Oh, I'm thinking," and there's a period of time in which you are completely identified with your thoughts. You are one and the same with them and you're completely at their mercy.

I've found more and more that if you are aware of whatever is happening as an appearance in consciousness and not as yourself, it can really change your relation to many important factors that affect performance. For example, a quickened heartbeat. If you just say, "Okay, my heart's beating faster," without getting stuck in this existential crisis of "I'm so nervous right now." Then the physiological response your body might be having in the moment can be interpreted very differently, and it doesn't result in this kind of cascade of thoughts that can then send you into an even deeper spiral of anguish. And I'm not exaggerating with the depth of the visceral response that people have when it comes to stage fright. I mean, it really can be this kind of extreme fight-or-flight response. And many people quit performing arts simply because it is too difficult to bear.

SB: It sounds like separating yourself from the experience of your own performance is important. This reminds me of the interoception literature, which I know you've read a bit about. The research suggests that if you have better interoception, which is very connected to mindfulness, you have a better self–other distinction. And this can be helpful in cases of empathizing with other people where actually you don't want to be so caught up in the other person's feelings that you're just completely swept up by them and kind of forget your own, but it also applies to your relationship with your thoughts or bodily feelings.

NN: I've found—and this is very, very personal—but, you know, I find that the more mindful I am, and the more I realize where this all comes from and the less I'm lost in notes, the way I move with the instrument, or the amount I need to move with the instrument, really decreases. I was never one to make faces or fling my arms around. It was always kind of against my nature. But it's less and less the more mindful I am because I feel that it's kind of like, getting so lost in the experience that you kind of lose the command center as well.

Some of my colleagues may disagree, but I don't think that in order to portray a sense of tragedy, one should oneself be in a state of anguish. Or that to portray joy, one should be overjoyed. I think that if one is too overwhelmed by emotion, one just stops listening. It's not useful because you have to produce something that will then make someone else, the listener, feel that. But as the producer of that experience, it doesn't help if you are overwhelmed by what you're doing.

Inevitably, when you listen to a recording, it doesn't really come across. You may be feeling something, but you're so overwhelmed by all this that you aren't listening, and then you aren't controlling what you're producing based on the aural feedback you're getting. There is a very complex process of adjustment and tweaking that happens, live, in the moment, and that's something for which a mindful control of all these various elements is really important and really useful. Losing that awareness and just enjoying it in a way that isn't conducive to that kind of focus I think hampers effective communication. So there is this sense in which you need to have this soberness, and I think that really helps in making sure you can really listen to what it is you're producing.

I think this can especially be a problem for pianists, because if you're a string player, for example, you have to carefully listen to what you're doing in order to know if you're actually producing the right notes in the first place. The response that your finger gets from the fingerboard isn't enough for you to be sure that you're playing in tune. Whereas on the piano you could be totally deaf but you could still know if you're playing the right notes because the kinesthetic response is so clear. You might not be playing them very nicely, but you know if you're playing the right or wrong notes. Therefore, pianists have a tendency really not to listen. So there is something about piano playing that is something of a danger in terms of not listening to oneself.

SB: Do you train other musicians in mindfulness?

NN: What I'm focused on now is consolidating all the things I've done on myself and creating a kind of systematic method that I can transmit to others to help other musicians with this, because I really think that it could help a lot. And simply bringing together my understanding of these various fields in sports psychology and neuroscience and mindfulness to bring together some sets of principles. Interestingly, the relationship of my meditation and mindfulness practice to performing has changed as the meditation practice itself has changed. And that is evolving, so I don't know what's going to happen next.

I'd be really interested to ask you, what brought you into this research, on mindfulness and interoception and psychedelics, and your experiences with this?

SB: Yeah, for sure. It is a long story, but at the moment I am especially interested in applying mindfulness and interoception to social situations. For example, social interoception and the body-based kind of way of connecting with yourself and others and this sort of dynamic back and forth, paying attention to what's going on in your body and paying attention to, you know, the gestures and facial expressions and words of someone else in a situation and how that impacts you. And then kind of going back and forth that way.

NN: Sounds like chamber music.

SB: [Laughs]

NN: Because that's a whole other dynamic, you know, playing with other people. And then there's this back and forth and response. And I think, you know, when I read about interoception, I think often about chamber music and this idea of making music together with other people and that what they do can have a physical effect on how you feel then, musically. So it's good to have mindfulness online in these moments for effective communication.

SB: Absolutely. Yeah, and then just the short version as far as psychedelics go, is that more and more of the research in psychedelic science is pointing toward an enhanced interoceptive awareness as one of the benefits of the psychedelic experience. So I'm interested in the body-based effects as opposed to this idea that we're just reforming mental pathways, which is happening too, but there's quite a lot of evidence pointing to the ties between memory, emotion, and the body and how they're all intertwined, and how the socialization of your bodily self, or your relational existence, is formed early in life. And psychedelics and mindfulness can both sort of rewire that, it seems.

NN: You know, what's interesting when I think back to some of the literature I've read, is that some people say one could think of consciousness as being essentially this feeling of having a body. That we are in a body, that's where consciousness comes from, and the place in the brain where we feel emotions is the same place that monitors bodily functions, and that's where consciousness comes from. And this is really fascinating. Do psychedelics strengthen the connection of, let's say, those midbrain parts to the prefrontal cortex or something like that? Are we more in touch with that? I don't know.

SB: It's an open question. I'm not sure about strengthening the connection, but they at least seem to loosen the screws that are in place so that they change your experience of your bodily self and therefore your psychological self.

NN: Do you practice mindfulness? Is it something that you're interested in, like personally or just from a research perspective?

SB: I am a Reiki practitioner. So I don't have a whole lot of experience in mindfulness and meditation practice, per se, but to me Reiki is one form of mindfulness. I would say it's most closely related to open monitoring because it's all about creating this nonjudgmental space where you are just receptive and paying attention to how your body feels.

NN: Open awareness is one of the most important mindfulness practices, I think, especially for people who are experiencing stress. If you just ask people, "Where are you feeling it and what is it like?" [it becomes clear that] they don't actually pay attention to what it feels like in the body. When they're suddenly asked to pinpoint it, it's like, "Oh, it's just this pattern of energy." And all of a sudden, everything that you thought was happening to you isn't there.

SB: My project is about mindfulness, but it's also about connectedness, which is a related concept. We connect more fully when we pay attention. And so I'm interested in whether musical performance helps you feel more connected, whether that's to yourself or to your audience or something larger than yourself.

NN: It's on many levels. First of all, I think that the sense of internal connectedness takes many forms, first as a purely physical one. If I don't play for a while, I feel like a drug addict because there is this physical addiction. Just sinking one's hands into the keys and that kind of response—that's definitely part of it. I don't think one can really practice for several hours a day without having that kind of connection. I do know there are those who practice to perform and don't really enjoy practicing. That's not the case with me. If I had no concerts, I would still practice because I just enjoy it so much. Then, of course, everything else that it does spiritually, if that's the word we should use.

It's very interesting because in the past couple of months, I've been performing without an audience, in livestreams. There's just the camera and the microphone, which is really like recording, but a live recording that goes straight to people listening. And it's very interesting. If I play a couple of times live on the BBC, for example, there may be hundreds of thousands of people listening in that moment, but I'm just in a room with the producer sitting there with a microphone, so at the same time there is this intimacy in the space of recording. That's a really surreal experience because it's disorienting. Who are you playing for? Are you playing for that one person at the microphone or are you playing for all those people listening at home? We're so used to having a very clear idea of who the audience is.

Interestingly enough, I've found you can still feel that sense of connection, even if you don't see who you're playing for. Which is strange because, first of all, to describe the live concert experience—I mean, that is very visceral. The nature of the audience also gives you a lot of energy when you're on stage. Pianists are really quite exceptional in this regard because we mostly play alone. Yes, we sometimes play chamber music and have lots of performances as soloists with the orchestra. But those of us who play solo a lot—we're just on this enormous stage alone in front of, let's say, two thousand people. And there is this very unusual state of consciousness one ends up with onstage. There is this total silence and total focus from everyone present on what you are doing, and you can almost hear that focus. It's very strange, but that silence is pregnant in a way that the silence of an empty hall wouldn't be. Even if no one is coughing or playing with their cough drop wrappers, you know that there is something going on there.

And there is a connection. Maybe especially in the silences, there is this common breathing together that one can really feel. And I think that what's interesting about the performing arts and music is that there are a couple of people producing what's going on, and then many people listening to it, and everyone's observing the same thing, but everyone will be receiving it and interpreting it in a very different way.

So there's something very interesting about that shared experience. There could be a Venn diagram for those two thousand people where all of this overlaps and where there are differences, and those differences will be due to people's experiences in life, their knowledge and familiarity with the material being presented, what they ate that day—all of this will affect how they interpret what is being presented to them by the artists onstage. I think that's a really remarkable thing. Even if the same person comes to hear the same piece played by the same performer on a different day, that experience will also be different. That's really one of the wonders of live performances. And somehow I think that . . . speaking to some of my colleagues who have also been doing a lot of these livestream performances during this time . . . we are able to feel something like that even when we're just playing for the camera in an empty concert hall.

You can somehow feel that it is being communicated to a broader audience. This may be purely a simulation in our heads, or the fact that we get some nice messages after we come off the stage, or whatever it is—but it does create the sense that one did connect with the audience. And, of course, a lot of the time it's a much larger audience than one would otherwise be playing for because more people can listen to something when it's being streamed digitally.

SB: It's interesting to think of all the different settings people could be listening from, cooking dinner at home or driving in their car or whatnot. It's like your music is being diffused into many different parts of a person's life, which means they're connecting with it in different ways than they would at a concert hall.

NN: Absolutely. At the same time, the concert hall is an interesting place because it's one of the few settings where your behavior is really going to affect the experience of other people. There is a lot of power in everyone being very attentive to what's sometimes even a very quiet sound coming off the stage, all together. It does shrink the size of the hall.

Not to get too technical, but there is this conversation we often have in performance technique about the question of projection—how to make sure that the sound you produce will reach the back of the hall and not get lost. And usually what that results in for most people is just not playing too quietly, always having a kind of minimal dynamic volume level so that people don't need to strain too much to hear what you're doing.

And I tend to not agree with this theory of projection. I think that a lot of it is about the intensity rather than sound volume. One can play extremely quietly, like really on the verge of audibility, but with an intensity that just forces everyone to be extremely quiet, and you can really feel that this huge space just contracts and shrinks. There is a sense in which somehow the artist onstage is able to manipulate the space. It's a really, really interesting idea. The space changes depending on what one is doing, and being aware of that is really special.

One very great pianist described being at Avery Fisher Hall, now David Geffen Hall, in New York. It's at Lincoln Center. It's a huge hall—there must be some twenty-five hundred seats in it. He was listening to Andrés Segovia, one of the very great guitar players of the 20th century. No amplification. The guitar is an extremely quiet instrument—it's very hard to get anywhere with it without amplification. He'd say that you could hear every note because everyone was so still, and it brought this huge hall together, just very concentrated in the experience of the single lone guitar player onstage. And I think those are some of the most special moments. There are very few situations where one could feel so close to so many thousands of people at the same time. It's a really unique experience and a very visceral one.

When I was 16, I had a boyfriend I was deeply in love with. We had to be apart for three weeks just after we started dating, when I went to the Caribbean with my parents, and it was torture. In our cabin near the jungle there were giant spiders and rats. I slept upstairs with mosquito netting all around me and a letter he'd written me folded next to my pillow. I listened to the mix CD he'd made me as I went to sleep every night, his black band T-shirt covering my small body. We were together for four years, and the love between us was one of the most powerful experiences of my life.

We don't speak anymore, and he is engaged, but his love revisited me the morning after things fell through with C. I was up at 6 a.m. on a Sunday, stirred by happenings from the previous night. I sat at my window, and for some reason I was moved to listen to a song my first boyfriend and I used to listen to together. I thought about that mosquito netting, and thought it odd that my heart shouldn't feel broken in that moment but rather the opposite: strengthened, whole, protected. It wasn't just recalling how he used to love me, or how I used to love him, that produced the glowing in my chest; it was the feeling that his heart was keeping mine company, now, in the present. The hardest thing about loving someone who doesn't love you back is that the love has no place to go, nothing to connect with. And suddenly he was there, loving with me, as I heard the song. The love had no recipient or direction; it was just happening.

In many North American Indigenous cultures, process and transformation, rather than fixed states of being, are the primary mechanisms of life. It follows that sound, vibration, and song are believed to be the creative, generative forces within the universe. The way events are experienced and communicated resists Indo-European language structures and Newtonian accounts of the physical world. "What is possible in Blackfoot may be impossible in English," says Leroy Little Bear, JD, a Blackfoot researcher and professor emeritus at the University of Lethbridge, in the book *Blackfoot Physics*.

James Sa'ke'j Youngblood Henderson, JD, an international human rights lawyer and member of the Chickasaw Nation, is struck by the many metaphors in English that refer to seeing. We say "Ah, I see" to mean we understand something, or speak of an "illuminating" idea. "We have perfectly good eyes," Sa'ke'j says. "If we want to discourse about chairs and tables we can simply point to them."

His language is not a duplication of sight, he says, but a complement to it, channeling a world of sounds and energies. Speaking is not meant to be a representation of reality, or something separate from it; it is meant to be an integral part.

"Some mornings I wake up with my head full of rhythms, and rhythms of rhythms, and rhythms of rhythms of rhythms," says Sa'ke'j, "and to have to speak English is like putting on a straitjacket."

In the 18th century, a Jesuit priest, translating an Algonquian dictionary, misinterpreted the term *Hipiskapigoka iagusit* as "a healer sings to a sick man." Two centuries later, modern linguists determined that it was actually a verb that expressed the act of singing and included, as modifiers, one who sang and one who received the song.

"What is really happening is singing," says holistic physicist and author F. David Peat, who spent the summer of 1980 with the Blackfoot tribe of Alberta, Canada. "The healer cannot really say that it is he who is singing; rather, the process of singing is going on. The singing is the primary reality, for it did not originate with either person, nor was the healing something that passed in a transitive way from one to the other. The singing sings itself. The healing heals."

We say life is the property of an object, that we "are" alive. We say we love a person, that someone is the object of our affection, as if the love originates from within us. But these energies can also be felt as independent forces, occupying a body—a verb, a process, a happening—transcending time and space and context. Life, embodied. Love, embodied.

The singing sings itself. The healing heals. The loving loves.

NARRATIVE

Solana Booth is a historical trauma and generational healing expert based in the Pacific Northwest, just south of Seattle. She uses Indigenous teachings—traditional and contemporary storytelling, diversity training, conflict resolution, and cultural presentations—to engender behavioral health and rehabilitation. She is the SeeQuilLouw program director at Advocates of Sacred, whose mission is to champion, integrate, and cultivate Indigenous healing modalities. She employs pre-perinatal psychology, somatic archaeology, generational brain spotting, birth and death work, traditional art, and her Positive Interconnectedness model. Booth is also a leadership trainer for healthcare professionals in historical trauma and family recovery intensives, a documentary filmmaker, advisor to Decriminalize Nature National, and executive leader at the American Psychedelics Practitioners Association.

As a traditional medicine keeper, Booth is in the process of opening the Recover Me in Wellness Center, with a focus on "Mother's Breath" (otherwise known as plant medicines), first foods, breastfeeding, canoe, and storytelling. As noted in her bio online, Solana is enrolled into the Nooksack Nation of the Chief Sam George Family from Beatrice Anderson and Samuel George, White Owl House of the Wolf Clan and Chief Joe Ortiz from Mohawk of Bay Quinte where the Peacekeeper was born. Her Paternal association is Tsymsyan of the Violet Atkinson and William Booth Family, Raven Clan. She is a mother of nine and grandmother of two baby girls.

Connecting over our shared roots in the Pacific Northwest, we spoke on a video call about relational health in the context of somatic archaeology and generational trauma, with a focus on storytelling.

SB: What would you say is the driving force behind your work at the moment?

Booth: What I'm calling Indigenous relational sense-making is the foundation of my work in everything that I do. It's almost like I can't help it. For me, it's a natural instinct to know that I'm related to Mother Earth, and that her breath—be it psychedelics or root medicines or birch bark—is these medicines, these plant teachers. When we talk about human nature, human nature is Mother Earth. She is our nature. She shows us all these things that we need to survive and thrive and relate. She's so on point that even her medicines—they just find us. Even in urban areas, where there are people suffering from liver disease or some kind of stress issue, there's a ton of dandelions outside their house. That's the medicine that they're needing. Usually the medicines that our bodies need end up popping up through the concrete or showing up in our yard. They show themselves to us all the time. That same visibility that we give them when we forage or harvest them is also what we need as a people. We need visibility. We need to be in relationship. We want that mirrored reflection, but we want to know how we can be in relationship. How come it's so uncomfortable? Why are people self-medicating with synthetics so much? It's about relationship. And Indigenous relational sense making is an all-inclusive, very dynamic, diverse lens to facilitate generational trauma recovery through somatic archaeology.

SB: What does somatic archaeology involve?

Booth: Somatic archaeology is a five-step, five-protocol process that allows the client to orient with their body, orient inside. Some things happen spontaneously, instantly, with us as a people. And our bodies only know how to survive; they only know how to thrive. They're so high tech on their own that the moment a traumatic experience happens, the moment something happens that makes someone hold their breath or frown or clench their ankles or tighten their jaw—the moment that happens, almost right before it happens, your body is like, "Oh, okay, it's that kind of thing." Because your body can sense things before your mind can catch it. Your body can sense things before your eyes identify it. Your body can sense it before you hear it.

So what happens when that thing makes you frown or hold your breath or tighten your ankle or clench your jaw is that your kidneys send off a hormone that contributes to the mechanism of the parathyroid (PTH) hormone, which goes up and down from your kidneys to your thyroid. And your thyroid says, "Oh, yeah, this is not good," or "Oh, it's kind of manageable." When we overwork our bodies, when we are living in a high stress, toxic environment, our PTH is no longer balanced—that signal from the kidney to the thyroid. And so now, to find the balance—because as soon as it comes from your thyroid and tells your body how much, if you don't have any because you've used it all, your body starts to take from your bones—you end up with osteoporosis or calcium deficiency or injuries, accidents, ailments, because of this degeneration.

The reason that I focus on the kidney part in explaining somatic archaeology is because women, typically, overall—like, 30 percent more than men—stress, get osteoporosis, do the heavy lifting. When we're doing the heavy lifting and our bones are getting robbed, we just get more and more sick. And how can we expect to model appropriate relationship in front of our children and our grandchildren or with our partners when our bodies are not functioning to their highest potential because of a toxic environment? So round and round we go.

SB: I was going to ask you about that, actually—the effect on children.

Because the interoception research suggests there's a link between suppressing your bodily signals in early development and mental health problems later in life.

Booth: The babies and the pregnant people are the ones I live and breathe for. Like, I could almost cry right now just thinking about all the nonsense that we put pregnant people through. And as an Indigenous nation, my heart is just broken. I've lectured and reached out and every time I'm storytelling in Indian community, I bring this up and say things like, it's unfortunate that we are treating our chanupas and our canoes—all these sacred instruments—better than our pregnant women. Our buffalo skulls or medicines, our cars or trucks. We're treating them better than our infants. We're treating them better than our toddlers. Our toddlers are running around in dirty diapers. But we're being extra careful around this canoe. That's not okay. These are just instruments. *These* are beings. The instruments are just a way to satiate the human need for that thing, that tangible thing, when it's already in us. We are the chanupa, we are the canoe. We're full of water. Our bodies carry that water just like the canoe.

My heart is just, like, achy big time. And I'm not blaming us. I'm just calling us back to action. I'm wanting to remind us exactly what we can be doing, and how we do it is irrelevant. We've just got to do it. We've got to make sure that our pregnant people are okay and our babies are fed and that they feel good and that they feel loved and that they know they're supposed to be here. Regardless of what happened in utero, regardless of what happened at birth, even if your mother surrendered you or if there was a rift in the family and now you don't have mom or dad. There're so many orphans, native and nonnative, who don't understand and don't know that Mother Earth is still here—your actual Mother Earth is still here. Your birth mother, the one that delivered you to here, that mother is not available. But you're here now. What are you going to do? Let's find a way to relate because you have to survive.

SB: And where do you see psychedelics entering this picture?

Booth: I really don't like the word *psychedelics* because it was by this white guy and I don't care to learn his name and remember it. That's how much I hate it. But of course, before a presentation, for example, I have to say those names and all that. But I do say that I want to change the name to Mother's Breath or something similar because I feel like it's people, the orphans, the abandoned, the rejected, the imprisoned, the ones who really take up and tax us with their needs. They tax us with their poverty and they tax us with their domestic violence and their drug abuse and their sexual assaults and all these things that come from a poor relationship. That's where it comes from. And I'm not blaming the mothers that we're not able to connect, that we're not able to connect their babies and have that positive, healthy attachment. I'm saying it's okay. Let's find ways, and I do have protocols to heal women in their wombs, in their eighth direction, their reproductive system. Let's find ways to heal them, if they want to, because they also have to want to. But let's focus on what we can do. We can help the babies, we can help the children. We can start talking and invoking and inducing this healing through relationship, through an actual healthy, sustainable relationship to Mother Earth, which is our human nature, which is Mother Earth. I don't know why people don't get it.

SB: Time to tell ourselves the full story.

Booth: That's what I do when I'm hunting. I see more evidence. I hear more stories of the hydromorphology of the creek, the glacier, mountain, to the creek, to the river . . . the stories of the stones along that river and all the different beings that are out there. Usually, the stones tell the river story, the river tells the salmon story, and so on. So all their stories, including my family, my lived experiences—which have been very crazy, awful, sad. It's been a struggle, with very proud moments to grow through all these things. So with all of that, it's about collectively and individually discerning: how can I make my work even more practical? How can I embody these truths and be honestly accepting, radically accepting of this reality? To stay radically humble?

My oldest daughter is an addict and she was sex trafficked at age 16. I caught (birth catch) her two daughters who are my granddaughters and I love them so much. But she's still struggling right now because she never was able to get help. She wasn't able to get recovery from the trafficking. And the trafficking happened. That's a generational trauma because that's what was happening since 555 years ago. And these things are not going to stop unless we interrupt the cycle. They're not going to stop. So maybe it'll skip a generation, but it's not going to stop. All these knee injuries and the hip injuries and the hip replacements and all these things are stories that were never finished, that were never reconciled. So my way of practicing somatic archaeology is a lot of somatic storytelling. Love won't stop either, ever. They are fighting for a sacred space.

SB: This is such a powerful use of narrative, the way you're talking about it. I hadn't thought of it that way before—it's about putting all these pieces together that look from the outset like separate pieces but they're all interrelated.

Booth: Yeah. Just like me and you and Mother Earth and even me and you in the Pacific Northwest.

SB: I know. I don't question these things anymore. If only a few more people felt this way. But I think we're getting there. It's so interesting reading about interoception because in some ways it's like a science of spirituality. Being connected to self is being connected to life. I'm wondering if neuroscience will start to include the body more and that will be one key to understanding.

Booth: I totally agree and I think that's what's encouraging, is this project that you're doing, because it's motivating me again. It feels like you're calling to action all these things, and clearly you're not the only one that's craving that kind of truth and that desire for the ultimate inner healing that's going to spontaneously start making all these things connect. Because generational trauma and pain can spontaneously hit us. You never know when you're going to get hurt or get into an accident or get a disease. It's always a surprise, always a complete shocker that makes time irrelevant. You lose track of time and that itself has a big effect. But it's the same thing when wisdom and knowledge hits us. It's the same thing when you get the "aha" and exhale and that healing starts and you cry or you're so happy something is happening and you get caught up and you lose track of time. You lose track of time because time is irrelevant to pain and love. Time is irrelevant to wisdom and knowledge and pain and trauma. And because time is irrelevant, that also means that these things exceed time. So you can heal generational trauma right now. You can heal 500 or 1,000 or 2,000 years of generational trauma. You can heal that spontaneously today.

Mother's Breath, the psychedelics, is the root medicine that helps us go into those places that are so dark and so deep and so brutal. Her breath can do that because that's what mothers can do. They created you. She just knows. She knows what our bodies need to be here. She knows what our bodies need to be in relationship. And I think that people don't lean on her enough for that kind of truth. People don't lean into her for that comfort, and that's really sad because they can. She's got some abundance to her.

SB: Is this related to your Positive Interconnectedness Model? Where does that fit in?

Booth: Oh my god, I haven't heard that in a long time. I don't get to focus on that a lot, not since the pandemic. But the Positive Interconnectedness Model is around healing somatic spoilage. What I call somatic spoilage is all the "isms," whether it's thought or felt or sensed or an actual physical action of racism, sexism, any kind of lateral violence, any kind of disdain or judgment. Not biases, but where there's a clear "Oh, no, this person is from *that* people." So even that little bit, that's the somatic spoilage inside of us that was passed down from generation to generation. That residue we're just born with, on top of not having a very good, strong emotional, spiritual, mental connection. I've had to do research and deeper work around: what are some ways that somatic spoilage shows up? And how can we start to interconnect ourselves to that spoilage so we can regenerate and recycle it back into medicine for healing? So we can change our minds, change our attitudes, change our language, change how we sense things, how we feel things to make it better for our children, make it better for each other. So that's where that model came from.

SB: That's amazing. Were you able to put it into practice?

Booth: I did. At the time, I was a board member for the Russell Family Foundation, and we started a Puyallup Watershed Initiative, and it was a $10 million, 10-year initiative, and we ended up making it $23 million and started our own nonprofit and used the Positive Interconnectedness Model to create a change theory for the entire community.

And that's also part of my drive for the work, with Mother's Breath—I would love for more tribes to get on the bandwagon. I would love for tribes to create their own canoe in this space and take their own agency so that they can not only profit but reorient themselves to these plants and reorient themselves to our actual ways and destigmatize the myths around these plants.

SB: Do you see some of these myths actively causing misunderstanding around plant medicine practices?

Booth: I do. It's very tricky to talk to folks about psychedelics or plant medicines in Indian Country because of a lot of stigma. But because of research being published with *Forbes*, CNN, and the mainstream media, people are softening. People are softening a little bit more and more. And again, I feel like we can really use storytelling to help not just soften but reorient. I believe this kind of reorientation to relationship and love to our plants and each other can be spontaneous and powerful. The same way that it was taken away—that fast—we can reconnect back to each other, back to ourselves, back to these plants.

> It was December, nearing the holidays, and I'd been seeking ways to integrate not only psychedelic experiences but experiences of heartbreak and estrangement—one's whole life, really. On a whim, I contacted the German psychologist Marc Wittmann, PhD, to ask a few questions about psychedelics. We ended up discussing time. He recommended a book by an esoteric called Maurice Nicoll.
>
> "Many of us, as a result of our reflections on life, must have come to the conclusion that something quite definite holds humanity back, not connected with commercial agreements or political questions," wrote Nicoll, MD, a British psychologist and colleague of Carl Jung, in *Living Time and the Integration of the Life*. "It cannot get beyond the stage in which it is, and keeps on turning in a circle. Civilization fails to get beyond a certain point. Some further growth of it is demanded which it seems incapable of undertaking, and from which it turns back . . . the whole question is about this higher level of consciousness and what will awaken it . . . nothing gained from books can possibly do this. It always remains a question of the *inner perception of ideas*."
>
> Though he never uses the term in the book, Nicoll's inner perception of ideas—contrasted with the process of collecting data from the external world outside our bodies, ranging from the light that reaches our eyes to the podcasts we listen to—is interoception. It's the immaterial, contrasted with the material; the invisible, contrasted with the visible.
>
> One way to awaken to it, he says, is to change one's sense of time, which can change the feeling of one's bodily self and the understanding of how to integrate one's whole life. Interoception researchers who also study psychedelics have pointed out the role of the body in time perception,

attributing both ordinary and altered time perception to the experience of one's internal state.

"Now, the mainstream neuroscientist would say: we have a representation of the world within the coordinates of time and space (the everyday sphere of ordinary consciousness); then we have a somewhat modified/disrupted representation of time and space during an altered state of consciousness, i.e., timelessness and selflessness," Wittmann told me in our email exchange. "We can nevertheless state that there are two modes of existence: the everyday and the exceptional. Both are valid. To some degree they can overlap. A person can change her personality over time by having exceptional experiences and integrating the feeling of being one with other persons and the world as well as the collapse of time into everyday living."

The key to doing this without psychedelics, Nicoll suggests, is to intuit a greater Time within which one's whole life is positioned; otherwise, one remains fully identified with one's life in passing-time, and whatever we remain fully identified with we can't integrate:

> Without the sense of the invisible there can be no unity, no integration—nothing but successive states, the ever-turning kaleidoscope of I's. For integration, ideas that halt time are necessary, and these ideas must feed us continually. And it is only through that particular kind of effort, whereby we realize our own invisibility, that such ideas can reach and feed us. Without this effort we fall into every moment, prone and lifeless, into the overwhelming stream of time and event, and the circle of our reactions. For at every moment we can sink down into our habitual state of consciousness—where no integration is possible . . . for then the sense of ourselves is derived only from the ever-changing response to the flicker of appearances . . . life carries us away, now up, then down. And the illusion of passing-time, and the thinking only in terms of time, causes us to fix our eyes always on tomorrow which never comes—for it is always tomorrow. So we live ahead of ourselves, strained out in time, and are never here, never in the place where we really are, the only place in which anything real can happen—in now.

That unchanging stillness within, if you can locate it, exists outside of passing-time and connects us to eternity. It's the one thing about the body that *never changes*, since it existed before it passed through us in this life and will continue to exist after this life. Outward truth cannot change us in ourselves because it "does not belong to us as inner experience and does not enter [us] from within."

There are different ways to harness Nicoll's inner perception to raise the bar for humanity, including through social connection. "The most important things are other people, and significant experiences," he writes.

> I believe that people who are really very significant to us are met with just when it is possible to meet with them—that is, when we are ready. If the life as a whole grows we may meet them earlier if this is possible—or as soon as it is possible. We must remember that there are different "times," or periods, on different scales, finally involving cosmic processes; and all this turning machinery of wheels within wheels must sometimes render things possible and sometimes impossible. If someone is significant to us, that person may or may not be influenced by this fact. But if there is a special understanding in common the influence must be mutual, and then the growth of one will be connected with the growth of the other. The inter-relation of these two people will not then be haphazard and accidental as is the relation of people in general. And in this significant inter-relation all the different aspects and possibilities of human relations must be thought of.

Were our consciousness attuned to this, he says, we would *see into* each moment and thus leave a trace of ourselves. "Only through some specially-guarded understanding can we increase the sense of the life as a whole, and so leave a trace."

MOVEMENT, YOGA, AND REIKI

New research out of the Max Planck Institute in Germany suggests that practicing "dyadic meditation"—where two people meditate together—may help us feel closer and more open with others. Similarly, research from the Virginia Commonwealth University School of Nursing found that the social connection aspect of yoga helps people manage depression. One person with whom I've both meditated and moved is Kit Kuksenok, PhD, a multidisciplinary artist, researcher, and transsexual who experiences queerness and transness as "the experience of being irreducibly a body." In their life and work, they explore the unseen interior of the human body's landscape using methods including, but not limited to, drawing, dance, participatory performance, speculative fiction, games, singing, contemplative breathwork and movement, corporate drag, autotheory, and experimental pedagogy. Together, we dove into the topic of movement and relational health.

SB: How does movement help you connect with yourself?

KK: Movement is a way to explore the body and that means first taking care of the body. When I stayed in Vancouver in 2015–2016, my commute was an hour and 37 minutes in one direction via bicycle, with a sizable hill. My eating, sleeping, and time-management patterns when I went there were suddenly—and surprisingly effortlessly—much healthier. These days I want to dance and move expressively, which means that in general, I work on flexibility, strength, and balance as a prerequisite for that kind of dance. This care makes the expressive exploration possible; the movement becomes a tool to explore the body and expand its range of possibility.

It's an oversimplification, but sometimes it's helpful to think of some things as being done *for* the body, and some things being done *with* the body: sometimes the movement is the tool (for exploration of the body) and sometimes the body is the tool (for achieving a particular movement). In practice, I think there's an interplay and a balance between these aspects in every movement practice, and no movement is purely one or the other. Nevertheless, it helps for me to think about that (is this being done *for* the body or *with* the body?) when I teach yoga asana, which is the postural practice of yoga, especially when teaching at the beginner level. Handstands are a good example of something that has a little bit of both: a desire to accomplish something specific and also the exploration of a new configuration. Someone who is new to yoga asana may be new to movement generally, or may be quite skilled at other kinds of movement, either *with* or *for* the body. They may have radically different motivations and attitudes around, in this example, practicing a handstand: to overcome fear, to have fun, to feel strong and graceful, to channel their energy in a particular way, because they can, or because they can't (yet).

A very strict and arbitrary time line around making "progress" with a specific movement makes the movement more *with* the body. A deliberate practice of checking in on what is available (in terms of strength, mobility, or concentration) before practicing makes the movement more *for* the body. Replacing movements with alternatives—or giving new things a playful try—without value judgment (such as practicing a handstand near a wall, without thinking of that as a deficiency) based on the outcome of such a check-in also makes the movement more *for* the body. I think in the context of teaching yoga asana, especially at a beginner level, it is my aim to demonstrate checking in, to offer alternatives, and hold a space where arbitrary goals or time lines can be questioned.

Though I think we all have different relationships with motivation and goals, I think I am at my best when I first create a space of total safety and acceptance for myself. From there, I find I spontaneously am moved to challenge myself and then very strenuous or unusual movement can be a mechanism of connecting to my body.

SB: How does movement help connect you with other people?

KK: Movement is primal communication, more essential than language. Something happens when I practice yoga asana with a dozen other people. We start to share a pattern of breathing. We share an energy: we can be playful and light—or we can be connected to a sacred experience together. Dance also provides a way to explore the physical space we occupy and to express or improvise narrative. Movement allows communication that does not need to resolve or to become definite; it can be both playful and sacred; both internal (to the body) and external (to the space, to others)—for me, the magic of this type of communication is suggested by [Mieko] Shiomi's < music for two players II > (1963). I think I'm drawn to the nonverbal communication of movement at least in part because my primary language is not my mother tongue, and I do not use my mother tongue in my daily life, so all language feels a little incomplete and a little removed from direct experience. Maybe this feeling is not so unique, because language necessarily abstracts and reduces, but in any case I think movement—dance, or physical theater/mime, or exercising in the same space, or hiking together, anything—is a direct experience of communication, and that can be really powerful.

SB: How does connecting with your body help you connect with others?

KK: Expression and empathy. Movement, particularly dance, can be used to tell stories and connect to stories outside of language. I am not a professional dancer, but dancing for fun helps me experience dance performance: by watching another body move, I can imagine my own body. It can be a kind of shared experience. Improvisation in dance is a very powerful connection experience where something is created outside of any individual bodies—between them. In movement forms like contact improvisation, there can literally be a center of gravity that the bodies place beyond themselves. These experiences are all completely outside language; but connecting with my body also expands my vocabulary of the body. Teaching yoga asana is also a fascinating experience because it challenges me to figure out how to describe movements through a mix of language and demonstration. Ultimately, both movement practice and watching movement demands direct experience; movement is irreducible to a description of itself.

SB: Has movement ever helped you manage mental illness?

KK: I think at this point most of my mental health management approach is movement. I was diagnosed with a mental illness more than a decade ago, and spent a couple of years on various psychotropic medications and in extensive therapy, which helped me to function, and helped me to establish much better routines, including exercise routines. As I became more involved not only in exercise—cycling, climbing, weights—but also in exploratory and expressive movement—yoga asana, dance—I gradually found I personally needed medication less. I do think therapy and medication were essential at the time, and I think I would be able to recognize if and when I might need those things again, but for the past few years movement has been a very effective mental illness management strategy for me. It is strange to remember how it was (back when I was diagnosed) to feel extremely alienated from my body, and also from my mind; both felt uncontrollable and unpredictable.

I do think a lot of social attitudes around the management of mental illness, chronic health conditions, and of neurodivergence can be very focused on gaining enough control of the mind and the body to be productive under capitalism. As Tricia Hersey has long argued, rest is an essential form of resistance and social justice, because the overwhelming pressure to control ourselves to squeeze out every available ounce of productivity not only breaks our bodies and minds, but also prevents people from having the energy or developing the knowledge and tools to fight for structural reform of unjust and inequitable systems. What is a better method for managing chronic illness: medication that makes me productive at work for a few years before my symptoms exacerbate to an intolerable degree? Or reducing the quantity or quality of my work to the extent possible to be able to take more naps (as Tricia Hersey, the Nap Minister, would urge), taking the same medication but without holding myself to the same standards of productivity I might have otherwise? What does it mean to "manage" a health condition that cannot be cured, or—in the case of neurodivergence—is only pathologized by specific social norms that are, in an increasingly more diverse and connected world, not necessary?

I believe it is important to interrogate every impulse to control the body or the mind, because often the impulse to be more productive in any given sense arises from an internalized neoliberal ethos. Movement practice, including yoga asana, is a simple opportunity to notice that desire to control and work with it. Why do I feel the need to do handstands today? How can I recalibrate that impulse into something that serves me as a whole being?

Although there is limited research as of yet on Reiki and mental health, it is currently being offered at well-reputed medical institutions around the world, including the Cleveland Clinic in Ohio. Even the discussions it has sparked around placebo effects have illuminated some useful findings on human biology. In late 2018, the *New York Times Magazine* reported on a group of scientists whose research suggests that "responsiveness to placebos, rather than a mere trick of the mind, can be traced to a complex series of measurable physiological reactions in the body; certain genetic makeups in patients even correlate with greater placebo response." Ted Kaptchuk, a Harvard Medical School professor and one of the lead researchers, theorizes that the placebo effect is, in the words of the *Times* article, "a biological response to an act of caring; that somehow the encounter itself calls forth healing and that the more intense and focused it is, the more healing it evokes." Simply directing your attention to another person's bodily self, in a caring way, seems to be a healing modality all its own.

I learned Reiki from Romana Cottee, a Reiki master living, practicing, and teaching in Berlin. We spoke about her relationship with Reiki and how it impacts her own life day to day.

SB: How does Reiki make you feel more connected to yourself, if at all?

RC: Having Reiki as a tool for daily self-care, like doing the self-treatments—and what I mean by that is usually the full self-treatment of doing three minutes and 12 positions each day—that noticeably shifts my energy. If I'm doing regular self-treatments, I find that emotionally I'm much more well-balanced. Not that I was unbalanced before, but I'm just able to observe things from a slightly more detached perspective: observe the emotion, but then just let it kind of pass through, and feel the emotion, but then don't hold onto emotions, like anger, for long. So, emotionally, helping with balance and just helping everything flow. Also, I think it helps to control unhealthy cravings. You're more in tune with your body and what your body needs, like food and not wanting sugar and that kind of stuff.

And it just helps me on a nervous system level because I'm a highly sensitive person (HSP), from the Elaine Aron book. I get easily stimulated, overstimulated by caffeine or noises or just stuff that's going on around me, probably the planets as well. So if I feel like my nervous system's triggered or overwhelmed, then if I do a regular self-treatment, it completely shifts how I feel. I feel way calmer. If I'm feeling tired as well—if it's more of a burn-out energy kind of tired—if I lay down for 45 minutes and do Reiki in the mid-afternoon, I feel totally refreshed afterward.

SB: Is that different from the effects of meditation, for you?

RC: Yeah, I would say meditation for me is a little bit more about the mind and observing the thoughts that come through my head. I'd say pure meditation is more of a mental focus of just having an awareness of thought patterns and an ability to shift and see them from another perspective. And Reiki does that as well, but then it does more, on a physical level. If I'm doing meditation and I'm not concentrating very well, maybe my meditation won't be very good. But if I'm doing Reiki, and my mind is a little distracted, the Reiki still works. So, yeah, something else is happening. I can sometimes try and meditate and not really be meditating because I'm distracted, but I could do Reiki and be distracted and afterward feel great still. And also when I'm meditating I don't necessarily feel as many sensations energy-wise as when I'm doing Reiki. I think there is definitely quite a lot of crossover and similarity, but they are not exactly the same.

SB: What about putting it in a social context? Does Reiki make you feel more connected to other people—whether it's in a session or outside of a session—just feeling more connected to people in the world or not?

RC: Not on a general social level, but doing Reiki for someone else or doing Reiki in a group with other Reiki practitioners like at a Reiki share . . . there's something deeply profound about it that brings me . . . not to tears like crying, but I feel it now . . . it's so beautiful that it makes my eyes water. Some things are so poignant or . . . I don't really know what the word is . . . I want to say raw . . . certain things happen that just trigger this response in us. It's so beautiful that you want to cry or something. And Reiki is like that for me. Like when I do the treatments for people . . . to be able to do that for someone, it feels very sacred and very beautiful. And also to teach people how to do Reiki and to do it with other people . . . it's a different level of like . . . it's like being in a temple or something. It's like you go into a holy place, like being in a slightly different reality, where it's more pure, less ego maybe, and just about love and compassion and recognizing the other and trying to support the other with their life, basically, by doing this practice.

But I think actually sometimes it makes me feel disconnected from other people, if they don't understand it. Like with dating, for example, I find that more difficult because my job is working with energy and I feel like . . . I don't know if it's just self-doubt or generalizations . . . but I feel like there's some kind of stigma maybe around people who do alternative healing. And I feel like if I make a dating profile on a dating app, I don't want to put on there that I do Reiki because I feel like I will be judged without people knowing me, like judged for doing something so "woo-woo," if you want to use that word. I find also that in some situations, people want you to justify what you're doing. [They say,] "Oh, it's not scientific" or whatever, even though actually I think it's very similar to a lot of things people are discovering with quantum physics. The science is catching up. But yeah, having these conversations with people . . . this is separating, I think. I definitely feel like there are some judgments around it. So it's slightly harder to connect sometimes in that respect.

SB: What about your connection to something larger than yourself?

RC: That's very, very strong, because Reiki is channeling healing energy or the source, and working with source. It's very difficult to find the right vocabulary, but Reiki is working with something outside of us . . . something that's within us, but also outside of us. And it's working through us and we're allowing it to work through us. So I feel deeply connected to something larger than myself. But in a way that's just . . . a knowing . . . I know that there's more and I know that it's real. It's not like I'm having a conversation when I'm doing Reiki and the Reiki God is telling me, "Now, move your hand here, now do this, for this person." It's just trust. You're connecting with source, but at the same time, you know that you're not able to actually be in conversation with it. You're working together without verbally communicating, if that makes sense. It's there, but there's still some distance. I mean, working with Reiki is way more of a connection to source that most people have on a daily basis. But you're still your own person, with your own mind, and it's not influencing us very much. You're completely yourself, but you're also connected to source at the same time.

SB: The way you described the knowing feeling . . . have you been able to transfer that intuition into other parts of your life?

RC: I mean, I think I was already very confident to trust intuition or myself or just sensing in the body since I've been traveling, since I lived in Australia eight years ago. Because I had all this empty space before me, like from going backpacking and not planning things. You just start to trust the universe or source to guide you every day to a new adventure and the right people and the right jobs, in the right places. And yeah, for at least the last eight years, I've been able to let go of worrying, about what's coming next and what I should do. So Reiki hasn't changed that for me personally because it was already there. But I definitely feel even more guided than I was before.

SB: Do you have thoughts on the mental health benefits of Reiki, like stories of using it to manage anxiety or depression or anything like that?

RC: I can't talk from my own perspective because my mental health is very good. I know it helps some people with anxiety. I think that it helps people a lot with their mental health. I have heard that from people. Generally with clients, I try not to keep all the details in my head, so I think it would be better to really do a study on that. But I'm sure that it does help people.

SB: Have you combined Reiki with psychedelics, or find the two related at all?

RC: I've used the Reiki symbols during a mushroom ceremony. I could feel the energy of them more, especially the Master symbol. My Reiki master said that when she did ayahuasca, she was feeling very intimidated and then she heard the voice say, "Use your Reiki," and it helped to move it through the body. I kind of also don't necessarily need Reiki when I'm on a mushroom journey because it's so powerful itself, and it's taking me through its own healing process. I had a Kambo treatment recently, which is not psychedelic, so I don't know if this counts, but it's still alternative medicine. It's a frog secretion from the back of a frog from the Amazon, and they burn the skin and then put the frog poison on there, and it triggers a really strong reaction in your immune system. You have a big cleansing purge. And I did use Reiki a bit, because you have to drink three liters of water just before you do it, so I did Reiki on the water. But psychedelics . . . I usually do Reiki on mushrooms before I take them, to connect with the mushrooms.

SB: That's an interesting way to think about using Reiki, in that context, because most of the time, people think about it as this tool for healing, but in that context you're actually using it as a tool to connect.

RC: Yeah, Reiki is this connection to source, and then plant medicine and psychedelics like mushrooms . . . they're so powerful and intelligent, it's like I'm connecting that source energy that flows through me to that thing. It's like an offering, and also like giving love to the mushrooms before I put them inside of myself.

> I didn't make major progress in terms of body trusting and interoceptive awareness until I started practicing Reiki in January 2020. Each Reiki training level requires a 21-day self-treatment afterward, advisable to complete before you continue practicing on others. So each night that January, I placed my hands over parts of my body, head to toe, and just felt as receptively as possible what was going on inside. It was therapeutic for a few reasons: I was honoring my body by making time to simply observe it without demanding anything from it, which is very different from going for a run or even practicing some types of yoga; I often felt inexplicable heat and tingling in different parts of my body, which made me pay attention to subtle signals without trying to predict or categorize them. I would sometimes use Reiki as a calming technique before a potentially stressful event, and saw that it worked not necessarily because it settled my nerves, but because it allowed for the possibility to arise that my body might be able to navigate any kind of discomfort I might feel and bring the situation back to homeostasis. That, in itself, is a calming thought.
>
> By the end of 2020, after completing the second level, I had started giving Reiki to friends and acquaintances and doing distance treatments. I find it especially therapeutic to give Reiki to others. It's an

intimate experience, requiring body trusting, social interoception, and encouraging the kind of receptivity and openness to possibility that psychedelics are often praised for. We let go of our defenses. We pay attention to nothing but each other's bodies—more specifically, the space between our bodies, the energetic essence of ourselves. Our eyes are closed, like two people blindfolded in a room together taking MDMA, turning on and tuning in. Afterward, our bodies hum with a quality that can only be felt, not spoken.

PSYCHEDELICS

One of the main reports from people experiencing the positive benefits of psychedelics, both in clinical and nonclinical settings, is the sense of "connectedness" and "unity" they feel with the world, themselves, and others. Where does this feeling come from, in the brain and body, and why is this particular emotion so powerful? How does connectedness heal us on a neurobiological level?

At the University of Chicago, Harriet de Wit, PhD, studies the physiological, subjective (i.e., mood-altering), and behavioral effects of drugs in healthy human volunteers. Current projects in her laboratory include 1) investigating individual differences in responses to psychoactive drugs, 2) effects of drugs at different phases of the menstrual cycle, and 3) effects of psychedelic-type drugs such as MDMA and LSD on mood and neural function. I spoke to her about her research on MDMA and sociality.

SB: How did your work lead you to studying psychedelic-type drugs?

HDW: I did a lot of work with amphetamine and methamphetamine, but really kind of to characterize their effects and to see how people differ in their responses to the drug in ways that might put them at risk for using the drug nonmedically. And so then it was a small step to go to MDMA, which resembles other amphetamines in many ways. Both drugs make people feel more social, both drugs increase feelings of sociability, and both drugs increase talking and interacting with other people.

But MDMA reportedly had this unique effect of making people feel more connected and closer to each other. And so my challenge as a basic scientist was to figure out whether there's something different about it, and what's different about it compared to other traditional stimulant drugs. So we started out a whole series of studies with MDMA to see how it differed.

SB: And what's different about it?

HDW: We looked at three categories of measures. One is how does a drug make people feel? How do they report feeling when they're under the influence of the drug? Another category is how they behave on tasks, in this case emotional tasks. And then the third category is physiological measures, so that might be like brain activity or heart rate, blood pressure, that kind of thing. We also measured emotional reactivity by looking at facial EMG, looking at smile muscles and frown muscles and things like that.

And then we asked a few questions that sort of differentiated the drugs a little bit. In one question—it was just a question somebody suggested—we used the adjective "playful." Are you feeling playful? And then, are you feeling loving? Amphetamine increased both of those adjectives to a moderate extent, but relatively speaking, MDMA did it more. So we were starting to see a little bit of a difference in how people felt on these two adjectives, "playful" and "loving." So that's reports of how they feel.

And then we went to a lot of more objective tests like recognizing emotions in other people. So we have a series of studies that showed that MDMA increases the threshold for being able to detect angry and fearful faces—so, negative emotions. You need more of the expression to be able to detect the emotion. And so that would fit with making it easier to interact with people since you don't feel like you're being judged. And so then you might be a little bit more open in interacting with them, either in a social setting or in a therapeutic setting. So that was one finding that we found consistently that I don't think we knew ahead of time. We didn't know that the drug would manifest itself in that way.

SB: So the reduced negative emotion may be just as important as enhanced positive emotion.

HDW: We've also done a test of what they call social exclusion. So it's a virtual ball-playing game called Cyber Ball. There are these characters—just stick characters—on the computer game, tossing the ball back and forth. In the early part of the task, you are one of the players and you are included. And then about halfway through the task, the other two characters throw the ball to each other and you are excluded. Surprisingly, [even] if it's stick characters on a computer screen, people feel excluded. So we wanted to know whether the MDMA would reduce that feeling of exclusion.

And it did. And methamphetamines didn't do that. So it seemed to have some effect on reduced feelings of rejection. So that was another piece of it we looked at.

There's also a measure of social touch. That's another thing that people in a naturalistic [setting] report, that there are alterations in how they experience sensory stimuli like touch. So there's a social touch task, where people rate a slow stroking as being pleasant and faster stroking on exactly the same place, with the same brush, people don't necessarily report as being very pleasant. So the slow one is pleasant, the faster one is not. So we wanted to know whether MDMA would increase the pleasantness of slow touch. And in fact it did, and amphetamine didn't. So again, another indication that this drug kind of specifically acts on something that has a social component.

So you can ask the question, "What is it about MDMA that makes it different?" One of the things that MDMA does is it increases blood levels of oxytocin. That's a hormone that is very involved in social bonding. And so some people speculate that some of these social improvements and social connectedness are related to increases in oxytocin. That's still controversial and people are still studying it. The drug also acts on the serotonin system, so it differs a bit from the other stimulant drugs because it acts more on serotonin. And serotonin is a neurotransmitter system that's involved in antidepressant effects. So that could also explain some of its antidepressant effects, even separately from oxytocin.

So those are some of the things that we've done with MDMA. It does seem to make people feel more connected with each other. It's being tested in clinical trials. I'm not really involved in treating patients at all. I do all my studies with healthy people. Obviously with PTSD, they've gone quite a long way.

Several weeks later, I spoke with Gül Dölen, a neurobiologist at the Johns Hopkins Center for Psychedelics and Consciousness Research. We discussed her 2019 *Nature* study, which showed that psychedelic drugs reopen a critical period in the brain when mice are sensitive to relearning the reward value of social behaviors. Based on this research, Dölen and her colleagues believe two things are required for psychedelics to be therapeutic in the context of social diseases like PTSD, addiction, and social anxiety:

1) the reopening of the critical period and 2) the right social context for the memory to be reshaped.

SB: Is it specifically the patient–therapist bond that allows psychedelic drugs to treat social diseases like PTSD?

GD: I think in a lot of cases, a lot of people—and especially the ones with PTSD that is very, very severe—are the ones that started with a traumatic experience in childhood during this maximum sensitivity to the social environment, which happens during the social critical period. So because of that, when a person is maximally sensitive to the social world, if they're injured by that social world, and the critical period closes, then that memory becomes an extremely well-ingrained worldview, and it's hard to dislodge it. And so I think that in the end of the *Nature* paper, we kind of ended with, "Oh, well, [psychedelics] might be just making the therapeutic alliance stronger," but based on this other more recent data and sort of thinking about it longer, I think that it's more than that. It's more than just the therapeutic alliance. It's about making those memories available to modification.

SB: How does this memory modification work exactly?

GD: The way I'm talking about it now is I call it "open-state engram modification." So you put the brain on MDMA in an open state where you're going to be sensitive to your social environment again, and then you either—through therapy or through processing your own memories or looking at photographs or journaling—what you're doing is bringing back the memory engram that is relevant to the trauma in this state where you are available to manipulate and make those memories malleable and rewrite them to respond to, you know, the realities of your current world.

So I think that when we think about what happens when someone has PTSD, for example, what we're dealing with is that during their childhood or youth, they are in a social environment and something bad happens to them, and in that moment, this response is very adaptive. They're protecting themselves by putting up walls, by guarding themselves from whatever is causing that injury. But over time, that adaptive response starts to become less and less and less adaptive until they reach adulthood and they're unable to form intimate relationships. They're unable to keep the job. They have a very negative view of themselves in terms of self-esteem, that they're deserving of love and being in the world, etc. And so I think the idea is—what we're doing with this, like with MDMA, is to go back and allow them to rewrite that memory in a way that's adaptive now that that traumatic event has been removed from their environment.

SB: And do you think that has to happen in a social setting, per se? I mean, I think it was your *Nature* paper that mentions that this phenomenon only happened when mice were with other mice. And, I mean, you could argue that that makes the case for the psychotherapist–patient bond. But of course, people have very transformational experiences taking MDMA on their own.

GD: I actually think that this is probably one of the most surprising and profound findings of the paper, is the setting dependence, because every other explanation that has been made of how these psychedelic drugs work from literally everybody else has always overlooked the fact that these experiences are very much modified by the set and setting, that they're context dependent. You know, it's not like people who have PTSD are going to raves and coming back cured. Yes, you can have profound experiences that are important in a therapeutic way outside of a doctor's office. But you're not going to have it if you spent the whole time just partying. In that case you're not engaging those memories.

And I actually think this speaks to a debate that's going on right now in psychedelics for therapy . . . that the pharmaceutical companies are really wedded to this idea that if we can understand the mechanisms of these drugs, on a pharmacological level, then eventually we can design a drug that activates whatever mechanism is curing depression or PTSD or whatever it is, and then we can design out all of those nasty psychedelic side effects. The psychedelic journey can be gone, right? Like, that's their dream. What every drug company wants. And then you have on the other side the psychologists who are like, "No, that can't be right because we know that we can achieve these psychedelic therapeutic effects even without the drug, as long as we can get them to this mystical place." And so we can do it with meditation, we can do it with a little bit of breathwork, etc. And furthermore, the stronger the psychedelic journey, the mystical experience, the strength of that mystical experience correlates with the strength of the therapeutic effects.

So these are the two sides of the debate. And I think our finding about the setting dependence of the ability of psychedelics to open the critical period kind of offers a middle ground between these two worldviews. So what it says is that the binding of the drug to the receptor opens a critical period. And that's the pharmacological effect that the drug companies have been so furiously searching for. Our hypothesis is that *that* is the mechanism. So any drug or any manipulation that can reopen the critical period has the potential for that therapeutic effect. But then on top of that, the setting dependence of it means to me that what the psychedelic journey is doing and the setting is doing is priming the brain so that the right memory and the right circuit is being brought into reactivation or made available for modification in this open state. And that's why I call it open-state engram modification.

And it's a middle ground between these two different views of how the thing is working. And I think that really solves a lot of the big questions. I think it says, "okay, mechanistically when we are evaluating a potential hypothesis or a new compound or a new way of doing these clinical trials, we need to address this issue of, you know, are we opening the critical period and are we effectively triggering the relevant engram?" Because if we're not doing either of those things, it's not going to work.

SB: I'm curious if you look at the interoceptive system at all or if you think it has a role to play in open-state engram modification?

GD: I think that this is actually a much older debate than this term *interoception* is. William James, in his *Principles of Psychology*, actually put out this idea that, instead of emotions being in the brain and the body just reacting to them, emotions are actually distributed through the body and that if you want to be in a good mood, for example, you should smile. For a long, long time in neuroscience, that idea was totally pooh-poohed—"No, no, the brain is everything and everything else is secondary." But in the last couple of decades, William James's original idea has come back into favor, especially with psychological experiments that have shown that if you don't tell people to smile, but you just make them hold a pencil in their mouth, for example, that this forcing of the smile muscles will trigger the engram, if you will, of happiness. And that engram of happiness is distributed. It's not just in the brain; it's in the whole body. We haven't tested it specifically, but my hunch is that reopening critical periods should certainly be able to modify memory circuits outside of the brain, including those that enable interoception.

A few months into the pandemic, I'd been hanging out with C in a tree by the lake and our pasts of drinking came up casually in conversation. He told me he'd gone to see a psychologist for a while but hadn't managed to change his lifestyle until moving to Berlin. Because he was vulnerable enough to share that with me, I shared with him that I'd remembered some traumatic experiences the previous year that I hadn't thought about in 20 years. I had never shared some of the things I said with anyone before. He responded with a kind of lightness, walking deftly along a line between showing sincere interest and expressing too much concern. When we got down from the tree, we paused at the edge of the lake and started joking around about something. Laughing, I stood looking out at the water. He stood close behind me and put his arms around me and just held me. It was a surprising feeling; I was a little dazed by it. He was responding to what I'd shared without speaking, just creating a container for my story with his body. Something about his nature as a person had allowed me to recall and communicate the experience without meaning to, and for the first time it felt manageable. The way he held me in that moment was exactly the thing someone should have done for me 20 years ago. I softened. I thought, "Oh, this is how you care for someone." And again, he approached it with lightness, as if to say, this is a very real thing I recognize in you, but also, there is a simple solution to what may feel like an unsuturable open wound. Just connect. With the right people, in this moment, through your body. Just connect. A few months later, the rare flower would be cut from my body. Because we laid these roots, it has since grown back in a different form.

IV. EXPANDING THE POSSIBILITY SPACE

Expanding the Possibility Space

AGENCY

From the mid-1980s to the early '90s, the American poet and activist Audre Lorde visited Berlin, teaching people how to change their bodies. Lorde's collection *Sister Outsider* was the first book I picked up when I moved to Berlin, long before I knew she had lived here, and it catalyzed my understanding of interoception.

Honoring the deepest part of yourself is the true meaning of the term *erotic*, she argued in her essay "Uses of the Erotic," and not the definition we've assigned to it since, which has been "relegated to the bedroom alone." Synonymous with life force, it is a creative power hidden in the subtlest intimations of self-expression.

"The erotic is a measure between the beginnings of our sense of self and the chaos of our strongest feeling," she wrote. "It is an internal sense of satisfaction to which, once we have experienced it, we know we can aspire. For having experienced the fullness of this depth of feeling and recognizing its power, in honor and self-respect we can require no less of ourselves."

The feeling potentiates joy, and also predates it:

> In the way my body stretches to music and opens into response, hearkening to its deepest rhythms, so every level upon which I sense also opens to the erotically satisfying experience, whether it is dancing, building a bookcase, writing a poem, examining an idea. That self-connection shared is a measure of the joy which I know myself to be capable of feeling, a reminder of my capacity for feeling. And that deep and irreplaceable knowledge of my capacity for joy comes to demand from all of my life that it be lived within the knowledge that such satisfaction is possible, and does not have to be called marriage, nor god, nor an afterlife.

To live outside in, however, can prevent us from realizing this satisfaction.

> When we live outside ourselves, and by that I mean on external directives only rather than from our internal knowledge and needs, when we live away from those erotic guides from within ourselves, then our lives are limited by external and alien forms, and we conform to the needs of a structure that is not based on human need, let alone an individual's.

Body trust allows us to return to ourselves, to know ourselves.

> We have come to distrust that power which rises from our deepest and non-rational knowledge . . . the considered phrase, "It feels right to me," acknowledges the strength of the erotic into a true knowledge, for what that means is the first and most powerful guiding light toward any understanding.

What's more, it's the birthplace of possibility.

I would read these words early in the morning, in 2018, as the sun emboldened the east-facing window of my shared flat. The night before, I might have had a glass of wine on my own somewhere, skipping dinner with my flatmates for reasons I didn't fully understand. Meeting Lorde on the page, these reasons swam inside a visceral space, making room for clarity and compassion and self-inclusion.

"There is a dark place within, where hidden and growing our true spirit rises, 'beautiful/ and tough as chestnut/stanchions against (y)our nightmare of weakness' and of impotence," she wrote in another essay, "Poetry Is Not a Luxury." "These places of possibility within ourselves are dark because they are ancient and hidden; they have survived and grown strong through that darkness. Within these deep places, each one of us holds an incredible reserve of creativity and power, of unexamined and unrecorded emotion and feeling."

This interoceptive, felt sense Lorde was describing, which can help a person move toward wellness, is the same felt sense that, when suppressed for long enough, not only precipitates individual disconnection but can lead to global crisis. Social justice movements such as transformative justice (TJ) step back to consider the ecosystem surrounding a crisis, and move toward healing by doing what's necessary to prevent further transgression: by pulling the harm up by its roots, as relationally as possible, without creating more harm in the process.

"What environment enabled the silencing to go on such that this pattern was able to continue until a crisis?" asks Esteban Kelly, a Philadelphia-based activist who cofounded the Anti-Oppression Resource & Training Alliance (AORTA), in a talk for the Barnard Center for Research on Women. What were the conditions of the community that allowed for this dismissal to be normalized, to imply that "it's not of a *scale* yet that we're going to intervene?"

In the case of systemic racism or sexism, he says, this might include subtle hints around white supremacy, patriarchy, and class power. The sentiment applies to everything from abuse within the family system to the etiology of mental disorder to climate change denial. But this isn't about pointing fingers so much as distributing the culpability by recognizing that we all play a part, however big or small, in the way things play out.

What are the conditions that prevent us from honoring something the first time we know it's true? Who determines statistical significance?

"In order to perpetuate itself, every oppression must corrupt or distort those various sources of power within the culture of the oppressed that can provide energy for change," wrote Lorde. In the case of transformative justice, state reform is seen as important and useful in reducing harm, but also runs the risk of exacerbating it, especially in the way Lorde describes. Instead, TJ focuses on community and local activism, recognizing more possibilities for transformation through community than through the state.

The tapping of resources we do ourselves, and within communities, becomes an expression of that very life force we thought we lost.

> Our acts against oppression become integral with self, motivated and empowered from within. In touch with the erotic, I become less willing to accept powerlessness, or those other-supplied states of being which are not native to me, such as resignation, despair, self-effacement, depression, self-denial. Recognizing the power of the erotic within our lives can give us the energy to pursue genuine change within our world.

Lorde's erotic is interoceptive awareness at its most powerful. She visited Berlin from 1984 to 1992, cultivating the Afro-German movement, until she died of cancer. She created space for possibility where oppression loomed, where there were few safe environments for Black individuals and communities to explore and define their own relational narratives. She invited white German women into her circle, and encouraged them to consider the ways they expanded or collapsed the possibility space for Black German women. She saw a naturopath who encouraged her to see her own wisdom as healing for her body. "Berlin put years on her life," recounted her partner, Gloria Joseph. These moments of holding Lorde's words between my hands became the impetus for this book.

"The white fathers told us: I think, therefore I am," she wrote, 40 years before consciousness research turned its gaze toward the body. "The Black mother within each of us—the poet—whispers in our dreams: I feel, therefore I can be free."

EXPANSION

Thanks in no small part to the quietest, most authentic, nonrational parts of internal experience, we have before us a new science of consciousness, alongside a growing collective awareness. "Consciousness," says the computational and cognitive neuroscientist Anil Seth, PhD, who helped popularize predictive coding, "has more to do with being alive than with being intelligent." In his book *Being You: A New Science of Consciousness* (2021), he argues against consciousness as a vehicle for thinking, communicating, and perceiving the world and instead proposes that these processes—and therefore consciousness itself—arose to support the physiological regulation of the body: "The most fundamental reason any organism has a brain is to help it stay alive." Philosopher and cognitive scientist Anna Ciaunica, PhD, takes this idea a step further, reminding us that we are alive within others before we even enter this world, in the womb of our mother. Our earliest experience of consciousness is therefore a relational one. The interoception research is leading us toward a science of the deepest, most personal parts of ourselves, which both defy and depend upon expression for survival. As psychologist Christian Jarrett, PhD, suggests, we're witnessing a convergence of "hard data and lived wisdom." Others, like Zen roshi John Daido Loori, are more cautious: "Mystery abhors naked exposure and explanation. What chance does a rational justification have against the beauty and mystery of our lives?"

> In 2021, after my initial round of interviews, I took the connection off-screen and went to the Netherlands. For a while, I'd had the intuition that I needed to spend time fully in nature, on a farm, with my hands in the earth. I contacted a few farm-stay hosts in different parts of the country but didn't hear back from anyone. One afternoon, sitting in a café, I looked out the window at a tree rustling in the wind inside a parking lot and seemed to register the movement of its leaves internally. Fingers poised over my keyboard, I typed a note to a Dutch psychedelic startup founder on LinkedIn, asking him if he knew anything about magic truffle farms, the commercial locations where psilocybin-containing truffles are produced to be sold in smart shops around the Netherlands. He told me to reach out to someone named Luc van Poelje, noting that: "Through him you'll probably end up at the right place." The same week, Sarit Hashkes, the cognitive scientist who had introduced me to the predictive coding theory of psychedelics, told me she was visiting Berlin and happened to be working in the Netherlands. Before long, I was organizing a month-long trip to visit several farms, write about a truffle retreat for war veterans, and connect with various people in the retreat and scientific communities. Farming itself quickly became a distant inspiration.

Since founding Psychedelic Insights in 2019, Luc van Poelje has facilitated truffle sessions for hundreds of private clients in the Netherlands, where magic truffles are legal. As a veteran himself, Luc recognized the need of an underserved group, and a need to serve them responsibly, the definition of which includes not only extensive screening but also creating the right supportive atmosphere. All of his guides have their own extensive experience with psychedelics, and many have military backgrounds, making it easier to connect with participants. This was Luc's first veteran truffle retreat with his team of guides, and the first of its kind in the Netherlands. It was also one of the first to incorporate scientific research, with an observational study conducted on-site by neuroscientist Luisa Prochazkova, a PhD candidate at the University of Leiden, investigating the effects of psilocybin truffles on quality of life and resilience in war vets with sub-threshold PTSD. All nine participants—one woman and eight men—were Dutch military veterans with moderate PTSD symptoms, short of a formal diagnosis. They filled out questionnaires before, during, and after the retreat on measures related to anxiety, depression, and other facets of well-being.

"Veterans are an at-risk group that has not been well studied since the 1950s, and now there are going to be two trials, one trial for sure in Leiden for people with proper PTSD where you come to the clinic and have an experience with a clinician on your own," Prochazkova told me. One of the important questions and warnings Prochazkova received from her colleagues pursuing clinical studies was, "'This is a very traumatized group and very likely to re-experience trauma from war and things like this, even if it might be dormant.' But I think the argument was always that there is a certain comradeship around veterans and the experience of them supporting each other and sharing their experience together is half of the job, and that is very special. The group dynamic helps people to share and be vulnerable in front of each other, especially coming from a highly militarized background."

Prochazkova accepted the project on the terms that she would document everything as objectively as possible. "I'm in the field of people who would never allow this to happen, but I said, 'Let's see. Let's see if it has any merit. I'm just going to keep my integrity.'"

As we spoke, eating pasta salad on the patio of a house in a forested area outside Amsterdam, I recalled my meetup with Michiel van Elk, PhD, a few days earlier. A cognitive neuroscientist, van Elk is the first scientist to have received a public grant from the Dutch government to study psychedelics.

"Most retreat organizers I've spoken to seem to have a sincere interest in science," he'd told me as we strolled side by side along the East Canal in Amsterdam with our paper cups of hot tea. "They are curious and want to know about the results and insights. But they are also anti-science, in a way, and anti-reductionist specifically. They are very much into spiritual realms and supernatural encounters and for them it's really like getting access to a different world. They talk about the wisdom of plant spirits and connection with Mother Earth, and they find us as researchers overly reductionist in our attempt to nail it down to questionnaires or brain data. It's indeed very neurocognitive and pharmacological because it is clear that if you block the serotonin receptor, no one experiences these effects, so it must have some basis in the brain somehow. But in the end, I'm very much open. I think science should also remain open and not end up in this reductionist [trend], while at the same time being realistic and saying this is the best paradigm we have right now."

And yet, what if . . . *the most fundamental reason any organism has a brain is to help it stay alive.*

After talking to Luisa, I spoke with several vets who had taken a 45-gram dose of truffles the day before my arrival. No one mentioned moments of military service as having come up during their trip, only insights about relationships with themselves and others. The connection between participants was palpable, too: something had been stripped away, not added, to the living room that morning as a sharing circle began around the hearth.

In the late afternoon, I sat on a bench outside the house with Ivan Calw, a Balinese-Dutch corporal who had served in the marines since he was 21, and spent time in Iraq as a child soldier. After initially being rejected from the special forces, he was finally accepted and spent 12 years there. He then injured his knee and could participate in neither the special forces nor his regular practice of the martial arts. "What is left of me," he asked himself, "if I can't do my work and my sport properly?" He went to his commander and asked for six months off to travel the world—to Nepal, India, the Philippines, and Bali, where he reconnected with his roots. "Because I couldn't fight, I was more into, 'How can I heal my body again?'" When he returned to the Netherlands, he felt that he no longer wanted to fight. "There's another path for me, the healing path." In addition to learning and teaching yoga, he began reading all the information he could find about psychedelics, until the research was no longer enough and he felt moved to find out for himself.

"So I went to a creepy house in Belgium and a woman opened the door and she looked like a witch with a long white dress and necklaces. It was scary as hell and they hugged me and it was the first time somebody hugged me that I don't know. After that, I drank ayahuasca seven different times and each time I got a little deeper—[into] my behavior, but also my pure essence. I saw my spirit, like I had wings—and how could I forget it—and it really inspired me. It was my own story but it inspired me."

I asked him about his experience at Luc's retreat, the day before.

"Because I already did ayahuasca and truffles many times, I already know where I got stuck. There is something in my body that wants to be seen, wants to be felt. It's stuck in my body."

SB: Where do you feel it?

IC: In my heart [squeezing his chest] and in my belly. In the emotional region and region for connection. I know this because I feel it in my heart. When I'm not connected to myself, it feels like [constriction]. Every time I take truffles or ayahuasca, it comes up.

[He spent some time with the feeling, lying on his bed, with a facilitator at his side.]

But it was enough for this moment and I went outside and stood with my feet on the ground looking at the trees moving and the plants, and I was like, "This is it, this is really it. Nature is the answer." I really felt that. You can read about it, hear stories about it, but this time I really felt the connection through my feet, through my whole body. It was like the earth was feeding my whole body, just standing on the ground and breathing.

The insight I got was, So if I take good care of you, Mother Earth, then she will give me back what I need. And I really felt this power in my body, in my hands. They got so warm. This energy is amazing.

SB: And that's connected to your pure essence? They are related?

IC: Yes, they work together. You can ask Nature for whatever you need. Somehow she provides for you the food and the nourishing to clean up your karma. It's like fixing an old story, and you need to be really well grounded to do that. I really think that's why we are here—to experience, but also heal ourselves from all these generations of war and pain. If I'm not the one who stops this for myself, I will give it to my children as well, and their children and their children. So I feel responsible for myself, for what I feel, and to deal with it and face it and work with it to connect with people. What we need, to do this, is the Earth.

SB: What about integration? How do you take these moments back into your daily life?

IC: What does it mean for the military? I don't know. I'm struggling to find my way, what I felt here and before, to integrate it into my job. I feel sometimes a bit different from my colleagues. You know, you can be in a situation and you can *be* the situation. For me, if there is a situation, I can step out of it and just watch it, have some distance, be calm with it and be okay. I just observe and make a decision by feeling, because I'm more in touch with my feeling, and from there I act.

SB: It comes from your body?

IC: From the body. And many people act from what they have learned through youth, school, jobs. But is it really the authentic you?

TRANSFORMATION

As I spoke with scientists and retreat organizers, trying to deepen my understanding of how everything was interwoven, I came, eventually, to the truffle itself.

Technically called a sclerotium, a magic truffle is the life force of a fungus. Legal in the Netherlands thanks only to a serendipitous loophole in the law banning magic mushrooms, it's the structure that stores nutrients and allows the fungus to withstand forest burns, droughts, and other ecological hardships. While gourmet truffles are reproductive structures meant for spreading spores around, sclerotia are vegetative structures that serve no reproductive purpose. They are masses of mycelium, the hard dark resting body meant to help the fungus survive. Depending on environmental conditions, they can unfurl underground into exploratory filaments or condense back into hardened masses. Many fungi with sclerotia, including morels, remain in this state for most of their life cycle. It's only when environmental conditions are just so that they signal mushrooms to grow. Burn morels, for example, only grow in the wake of forest fires, when an area of land has been completely decimated. Mushroom hunters have known this for centuries: in the 1800s, morel-crazed Germans would deliberately burn stands of forest just to trigger the mushrooms' growth. What catalyzes the prolific fruiting after fire is still an ecological mystery.

One thing that makes morels notoriously difficult to cultivate is the very fact that their life cycle includes the sclerotium. In the 1980s, Ronald Ower, a graduate student at San Francisco State University, finally found a way to grow morels in a lab using—as legend has it—a compound that mycologist Paul Stamets had developed for growing psilocybin mushrooms.

What signal could these compounds be sending, within a fungus, to engender such growth? Could it shed light on their effect in humans? Perhaps it's worth taking a wider, more ecological view of the question.

In 2018, Jennifer Frazer wrote a piece for *Scientific American* touching on the evolution of psilocybin. In the process of searching for psilocybin-producing genes in a population of wood decay fungi, she says, scientists made a discovery: there was less variation in the gene content of distantly related wood decay fungi than between decay fungi and their close relatives in other habitats. In other words: distantly related psilocybin fungi that shared the same environment had less gene variability than closely related psilocybin fungi in different environments. "The fact that their shared environment seems to be a stronger driver of gene content than shared ancestry," she observed, "is quite stunning."

LSD, which is derived from a fungus, plays a part in this story as well. When rye is exposed to the fungus *Claviceps purpurea*, the fungus grows upward within the stalk of the plant, eventually pushing forth through the head of the stalk and replacing what would be a rye kernel with a dark plug of its own design, called the ergot. This structure replacing the kernel isn't just general fungal material; it's a sclerotium. In some infected plants, the sclerotium drops off the head of the stalk and lies dormant in the soil until conditions become favorable for growth. Unlike psilocybin fungi, ergot is semi-toxic to humans unless synthesized into LSD or other compounds. It was used for medical purposes as early as the 16th century: for precipitating childbirth and stopping prenatal bleeding. However, "ergotism" (also known as St. Anthony's fire) became a serious condition in humans and animals in the 17th century, causing convulsions, hallucinations, or gangrene-like symptoms, depending on geographical location. Ergot alkaloids are used to some degree in medications today, especially for migraines and cluster headaches. But it seems to only work in "states of emergency" in which an intervention is required: ergot for childbirth works during labor but poses serious risks outside that window; ergot for headaches only works if the headache has already begun.

The sclerotium—the hidden life force of the fungus—seems to be a catalyst for pivoting from crisis to resolution.

With a little digging, you see flashes of this mystery reflected in the science, including in some of the leading theories on how classic psychedelics work in the brain. Pivotal states theory, introduced by Robin Carhart-Harris and colleagues in 2020, suggests that classic psychedelics activate a sort of emergency setting in the brain, meant for adaptation during times of hardship: "Psychedelics hijack the same neurochemical mechanisms that are engaged during, and likely exist for, situations where a hyper-plastic state and associated psychological change is felt as needed." Although psychedelics can produce this effect, the researchers argue that "the capacity for pivotal mental states is an inherent property of the human brain itself."

Posttraumatic stress and posttraumatic growth, they write, may be two sides of the same coin—diverging paths forward from a common environmental trigger. "We propose that the mechanisms underlying pivotal mental states have evolved to aid rapid and deep learning in situations of perceived or actual existential threat or crisis for the ultimate purpose of catalyzing psychological change when circumstances demand this."

Psychological change, but also, perhaps, bodily change. Relational change. Ecological change.

In other words, to keep what's living *alive*.

> When I returned home to Berlin, a friend said, "I thought you went over there to spend time in nature." But I had already connected to nature the moment I decided to go.
>
> One year later, following the same line of feeling, I would return to the environment I'd been raised in after being away for four years. The shape of the wound appeared unexpectedly and with unprecedented clarity, a familiar and ancient heaviness inside that weighed me down as the current of my family's life updates carried us swiftly along.
>
> "I want us to do something together," I said. Something inside me wants to be seen, I explained, and it doesn't require language in this moment, only collective attention. Sitting at a table with a dark river stone in my hand, I asked them to sit in the chairs on either side of me, hold their hands over mine, and attend to the stone. We closed our eyes and focused on it. Something rearranged and integrated itself inside me. When I opened my eyes, their heads were still bowed toward it, honoring it for bringing me to this place.

V. PRACTICES

Practices

Interoception is a growing research field, and there is still much to uncover, but we certainly know enough to begin putting some of these ideas into practice. Interoceptive awareness scales exist, with the Multidimensional Assessment of Interoceptive Awareness (MAIA) Scale, developed by Wolf Mehling and colleagues, being the most commonly used. You can take the MAIA Scale test yourself below or online at https://osher.ucsf.edu/research/maia. For each statement, answer 1–5 depending on how strongly you agree with it. You'll want to reverse-score (RS) items 5–10 on Not-Distracting and 11, 12, and 15 on Not-Worrying. Take the average of your answers to get your score.

Following the MAIA Scale, I've included an Interoceptive Connection Toolkit that combines practices from published researchers and practitioners with my own (nonclinical) recommendations.

MAIA SCALE: MULTIDIMENSIONAL ASSESSMENT OF INTEROCEPTIVE AWARENESS

Noticing

1. When I am tense I notice where the tension is located in my body.
2. I notice when I am uncomfortable in my body.
3. I notice where in my body I am comfortable.
4. I notice changes in my breathing, such as whether it slows down or speeds up.

Not-Distracting

5. I ignore physical tension or discomfort until they become more severe. (RS)
6. I distract myself from sensations of discomfort. (RS)
7. When I feel pain or discomfort, I try to power through it. (RS)
8. I try to ignore pain. (RS)
9. I push feelings of discomfort away by focusing on something. (RS)
10. When I feel unpleasant body sensations, I occupy myself with something else so I don't have to feel them. (RS)

Not-Worrying

11. When I feel physical pain, I become upset. (RS)
12. I start to worry that something is wrong if I feel any discomfort. (RS)
13. I can notice an unpleasant body sensation without worrying about it.
14. I can stay calm and not worry when I have feelings of discomfort or pain.
15. When I am in discomfort or pain I can't get it out of my mind. (RS)

Attention Regulation

16. I can pay attention to my breath without being distracted by things happening around me.
17. I can maintain awareness of my inner bodily sensations even when there is a lot going on around me.
18. When I am in conversation with someone, I can pay attention to my posture.
19. I can return awareness to my body if I am distracted.
20. I can refocus my attention from thinking to sensing my body.
21. I can maintain awareness of my whole body even when a part of me is in pain or discomfort.
22. I am able to consciously focus on my body as a whole.

Emotional Awareness

23. I notice how my body changes when I am angry.
24. When something is wrong in my life I can feel it in my body.
25. I notice that my body feels different after a peaceful experience.
26. I notice that my breathing becomes free and easy when I feel comfortable.
27. I notice how my body changes when I feel happy/joyful.

Self-Regulation

28. When I feel overwhelmed I can find a calm place inside.
29. When I bring awareness to my body I feel a sense of calm.
30. I can use my breath to reduce tension.
31. When I am caught up in thoughts, I can calm my mind by focusing on my body/breathing.

Body Listening

32. I listen for information from my body about my emotional state.
33. When I am upset, I take time to explore how my body feels.
34. I listen to my body to inform me about what to do.

Body Trusting

35. I am at home in my body.
36. I feel my body is a safe place.
37. I trust my body sensations.

This scale has been translated into more than 20 languages, including German, Turkish, Arabic, Mandarin, and French. It has been clinically validated and continues to reveal important insights about various disorders.

Body trusting, in particular, is emerging as a clinically relevant subscale. In his own research, Andy Arnold found low body trust to be the biggest predictor of loneliness among a population of students at the University of California, San Diego. It is also closely tied to eating disorders. "I find it pretty impressive that the three items hang together consistently across different populations, predicting clinical outcomes like the eating disorder literature that has recently been amassing on this topic," he told me.

People with depressive symptoms suffer from low body trust as well. "Lack of body trust appears important for understanding how individuals with depression interpret or respond to interoceptive stimuli and may represent the leading edge of interoceptive dysregulation seen in depressive disorders," write a team of researchers at Harvard Medical School. Their study also showed an inverse relationship between depression and the Noticing and Listening subscales.

The relevance of the other subscales are fairly self-explanatory. Self-Regulation and Emotional Awareness relate to mood disorders, while Not-Worrying relates to anxiety. A low score in Not-Distracting has been correlated with suicide ideation, though the same study found that suicide ideators are no worse at Noticing than non-ideators. One lesson from the clinical literature is that awareness of bodily sensations is not, in and of itself, correlated with well-being. How one *relates* to one's sensations—e.g., trusting, not-distracting, etc.—is what makes the difference. Still, one must notice in order to relate.

INTEROCEPTIVE CONNECTION TOOLKIT

A common concept within Tibetan, Chinese, and Indian medicine is that of the "subtle body," represented as channels, meridians, and "energy centers" (Sanskrit: chakras) through which "subtle energies" pass. These energy centers are believed to have an effect upon the mind, and researchers are beginning to link their role in contemplative practice to the science of interoception.

"Every mental event—that is, all states of consciousness—are said to ride the 'steed of wind' or 'energy' currents," write Norman Farb and colleagues. "It is currently unclear how these conceptualizations map onto scientific approaches of interoception. However, these concepts do suggest that attention to embodied experience is significant for self-representation and well-being, and therefore supports the more general hypothesis that over-dependence on top-down, or merely conceptual (in contrast to sensory) awareness significantly limits a human being's potential for relating to self, others, and the world."

The purpose of this toolkit, inspired by the interoception research literature, is to overcome those limitations.

ATTENTIONAL STRATEGIES

Interoceptive awareness may sound a lot like mindfulness. They are not exactly the same construct, but they have one primary thing in common: attention regulation. "Attention regulation is the basis of all meditative techniques and appears to be a prerequisite for the other beneficial mechanisms to take place," writes Jonathan Gibson, PhD, assistant professor of psychology in the Department of Humanities and Social Sciences at South Dakota School of Mines and Technology. The same is true for interoceptive awareness.

Differences arise in the way each practice approaches attention regulation. "For example," Gibson continues, "mindfulness does not often distinguish between attention directed to interoceptive sensations, exteroceptive stimuli, or conscious thoughts. This may be significant as several recent studies highlight that different types of attention elicit different neural signatures." Likewise, he says, the interoception literature establishes interoceptive sensibility (IS) as a facet of interoceptive awareness but often fails, for example, to "differentiate anxiety or hypervigilant style from a more mindful and open style toward interoceptive sensations. . . . Training individuals to focus solely on interoceptive sensations does not automatically imbue participants with knowledge on how to alter attentional style or mental habits commonly employed to avoid unpleasant sensations when those emerge."

The key to well-being through these practices is not just paying attention—it's *how* you pay attention. "There are a number of ways in which one can attend to the body, and each style can reveal different insights and understandings," Gibson explains. "Different types of attention can function like different focal points, each revealing certain dimensions that may be unavailable from other attentional styles."

FOCUSED ATTENTION: RHYTHMIC BREATHING, BODY SCAN, YOGA

"Focused attention on the body activates the insula and interoceptive network, which increases interoceptive awareness (IA) with its associated functions," Gibson writes. "Increased IA is closely related to changes in perspective on the self, which is consistent with Buddhist philosophy and the mindfulness literature."

Breath regulation, including breath work, is a type of focused attention. Several studies have found that focusing on the breath activates the interoceptive network. It also increases vagal tone, dampens sympathetic nervous system activity, promotes parasympathetic dominance, and facilitates stress regulation and cognitive control.

Body scans can be similarly therapeutic, but depending on your existing relationship to your body, you may run the risk of experiencing even more discomfort by paying closer attention. In fact, focused breathing has been found to be more consistently beneficial than body scanning, perhaps because it provides a safe attentional focal point (the breath) that body scanning does not. For this reason, body scanning is most beneficial when combined with a mindful attentional style.

During a study on Mindfulness Based Stress Reduction, Carmody and Baer found that yoga was significantly associated with changes in mindfulness—specifically observing, acting with awareness, nonjudging, and nonreactivity as well as improvements in well-being, perceived stress levels, and several types of psychological symptoms. Gibson suggests this may be a function of something more specific than mindfulness itself: increased interoceptive awareness as evidenced by neuroplasticity changes in the insula.

OPEN MONITORING

In contrast to focused attention, which simply encourages you to notice bodily sensations, open monitoring incorporates an appraisal component. It involves being open, accepting, and nonjudgmental about sensations without getting carried away. In other words, it proposes a *style* of paying attention, a way of thinking about what you are feeling, which is interoceptive sensibility in a nutshell.

This can prove especially useful, Gibson says, to those within a clinical population.

> When individuals with a history of abuse or trauma recognize sensory signals from the body, which triggers an emotional response, the space created by a mindful attentional style allows the individual to maintain awareness of their bodies instead of dissociating from those sensations into habitual or conditioned responses. Over time, participants can discover that their bodies can be a helpful resource rather than a source of threat that should be avoided. Those bodily sensations, which were previously encoded as a threat, can now be integrated into broader states of consciousness that can help the person develop a new self-schema and sense of safety in the world.

In this way, open monitoring can help people develop a meta-cognitive awareness that makes it easier to disidentify from their own emotions and feelings and simply observe them. "It is a recognition that one is having an experience rather than 'being' the experience," Gibson explains. "Bodily sensations can simply be experienced rather than transform the experience into a self-defining attribute."

Maintaining attention, as opposed to just quickly noticing, is a key part of this process. "There is evidence that the ability to maintain attention on conditioned responses appears to be necessary for successful extinction of conditioned responses." Open monitoring, in particular, supports this ability.

Compared to focused attention, which may just elicit discomforting sensations without resolution, open monitoring offers a potentially safer focal point from which to explore your embodied self, not unlike MDMA.

FOCUSING

Focusing, which is distinct from focused attention, involves "knowing" with the body and developing one's "felt sense." A felt sense can be a vague physical feeling, or an image, or something you notice in your body that gives you unique and relevant information about what's happening for you in the moment. Developed by University of Chicago psychotherapist Eugene Gendlin, PhD, in the 1950s, focusing developed into a six-step process based on his observation, over the course of 15 years, that nearly all successful clients in psychiatric settings learned to intuitively focus inside on this very subtle internal bodily sense that helped them resolve their dilemmas.

"Rather than attending to the body in an open-monitoring, accepting, non-reactive attitude, focusing attends to the body with a particular type of attention meant to unveil unconscious bodily knowing," Gibson writes. "This attentional style is not focused on mere body sensation or getting in touch with one's feelings, or simply observing them, but rather it is focused on connecting to a broader, deeper physical sense of meaning that is done by asking particular questions and waiting for the body to respond. This attentional style can open an aperture to a deeper dimension where sensations reveal a rich array of information all within a broader state of consciousness."

In focusing-oriented psychotherapy, the therapist asks questions and listens in a way that helps the client connect with their felt experience. You can also practice this on your own, asking yourself things like "Does this feel right?" and "Why don't I feel completely comfortable?" and "What is my body telling me to do?"

FLEXIBLE SWITCHING

While the above practices are useful for developing greater interoceptive awareness in private, or with a therapist, they don't offer much in the way of staying connected during everyday social interactions. How do you use interoception during a social exchange? Attentional switching (or flexible switching) is one of very few methods outlined in the current research literature.

"High-quality social connection is acquired and maintained by the *flexible use of interoceptive signals during social interaction*, which we have previously referred to as an aspect of 'social interoception,'" write Dobkins and Arnold. "The balance is about flexible alternation between conscious external attention [paying attention to the other person] and interoception, so that the internal signals can be appropriately linked with the external information."

As an example, Dobkins says that when conveying one's inner experience to another, it helps to name the objective event followed by how you felt, for instance: "When you said, 'I'm really busy and can't talk,' I felt hurt and abandoned."

> Interoceptive awareness is necessary, but not sufficient, for high-quality social connection, especially in socially challenging situations. Awareness needs to switch between internal and external experience, in a moment-by-moment fashion, so that the two may be integrated for adaptive learning in social situations.

In other words, the better you get at switching, the more quickly you learn what internal and external cues mean in relation to each other, which can help you manage your relationships.

MINDFUL AWARENESS IN BODY-ORIENTED THERAPY (MABT)

Developed in the 1980s by Cynthia Price, PhD, director of Seattle's Center for Mindful Body Awareness, Mindful Awareness in Body-Oriented Therapy (MABT) comprises a three-part approach to interoceptive training: identify, access, appraise. "While MABT and other mindfulness approaches involve both bottom-up and top-down processes," Price explains, "MABT is unique in its strong focus on bottom-up learning processes involving a focus on sensation guided by the use of touch to support learning interoceptive awareness."

1. **Identify: Become aware of and name body sensations.** *Body literacy* is the ability to identify and articulate bodily sensations. In MABT, it is taught by applying gentle physical pressure to an individual's body (e.g., shoulder), either through self-touch or touch by the therapist. The individual is asked to describe their own response to the touch, and to find words for the sensation. The therapist can provide a list of words as prompts if the individual struggles to articulate their feelings.
2. **Access: Breath flow; tissue softening; internal body attention.** This step is intended to guide attention toward internal sensation. The therapist encourages the individual to notice the feeling of breathing, the softening of tissue during a massage session, and finally specific areas of the internal body while at rest. "These various exercises often become well-used strategies for self-care that are incorporated into daily life to facilitate self-care and regulation," Price explains.
3. **Appraise: Sustain awareness; notice internal shifts; reappraise.** "It is in the state of sustained mindful attention that individuals most commonly experience new awareness or insight about themselves or a situation," Price writes. "Insight is understood as a change in consciousness that includes a shift in understanding, a psychological process thought to inform well-being in meditation practice. Such shifts in self-understanding often include new awareness of the links between physical and emotional sensations, involving metacognitive awareness processes that underlie cognitive appraisal of bodily experiences, and appear to be critically important for insight, integration of interoceptive experience into self-understanding (i.e., sense-of-self), and the ability to better regulate emotion."

One way of understanding the powerful effects of psychedelic experiences, especially ego-dissolution-level experiences, might be through the lens of sustained awareness. When the ego dissolves, so to speak, we no longer have a choice but to face what we're feeling.

lens of sustained awareness. When the ego dissolves, so to speak, we no longer have a choice but to face what we're feeling.

Following MABT, sustained awareness and awareness of internal shifts ultimately leads to cognitive reappraisal. Due to heightened interoceptive awareness, our interpretation of an experience might change, for example an experience we initially considered stressful that we now see as an opportunity for growth. "Developing the capacity for interoceptive awareness is thought to facilitate positive and adaptive reappraisal processes, a critical aspect of emotion regulation."

It's worth noting here that some studies have shown that people who named emotions prior to reappraising them reported feeling worse than those who did not name them first. "Instead of facilitating emotion regulation via reappraisal or acceptance," writes study author and psychologist Erik C. Nook, PhD, "constructing an instance of a specific emotion category by giving it a name may 'crystalize' one's affective experience and make it more resistant to modification."

RESPIRATORY PRACTICES

"Respiration is unique compared to other sensations (such as the gastrointestinal one) insofar as conscious regulation can immediately impact respiratory processes, and respiratory processes can affect emotion and cognition," write Helen Weng, Jack Feldman, and colleagues in a recent article, "Interventions and Manipulations of Interoception."

Slow breathing, in particular, benefits the sympathetic nervous system (SNS) by reducing reactivity and overactivation. Studies have shown slow breathing to be effective in reducing PTSD symptom severity. It may also reduce blood pressure and help prevent cardiovascular disease as well as chronic kidney disease, which is marked by chronic overactivation of the SNS.

Crucially, these studies have shown that slow breathing doesn't reduce SNS overactivation without a "mindfulness component"—that is, patients must be aware of their own breathing, and deliberately draw their attention to it, for these effects to occur. In this way, therapeutic respiratory practices both require and enhance interoceptive awareness.

ALIGNING DIMENSIONS OF INTEROCEPTIVE EXPERIENCE (ADIE)

Heartbeat detection tasks are still the most frequently used interoceptive training tasks in clinical settings. Developed in a clinical setting by Sarah Garfinkel, PhD, and colleagues, ADIE is a novel therapy combining two modified heartbeat detection tasks with performance feedback and physical activity manipulation to transiently increase cardiac arousal. ADIE was found to reduce anxiety in autistic adults specifically through interoceptive awareness training, "putatively improving regulatory control over internal stimuli," the authors write. "With little reliance on language and emotional insight, ADIE may constitute an inclusive intervention."

Some labs are developing products to help people improve interoception through heartbeat detection. Doppel, a wearable device, delivers a heartbeat-like tactile stimulation on the user's wrist. When delivered at a slow, steady pace, the beat has a calming effect, and has been shown to relax users before public speaking events. Other interoceptive technologies are sure to be on the way.

SOMATIC PRACTICES

General somatic practices—although they are just starting to be scientifically linked to interoception—very likely improve interoceptive awareness.

FELDENKRAIS

Developed during the mid 20th century as an exercise therapy, Feldenkrais involves directing attention to habitual movement patterns and uses slow repetition to introduce new, more beneficial habits. This somatic method, which focuses on enhancing brain–body communication, has been shown to improve interoceptive processes and psychological well-being in several populations of study participants, including female adolescent ballet dancers.

ALEXANDER METHOD

The Alexander method, often used in actor training to help actors perform more naturally, is an alternative therapy that focuses on improving posture and balanced use of the vocal tract. In a *Body Learning* podcast, Alexander technique instructor Imogen Ragone and trauma awareness activist Shay Seaborne discuss how awareness of bodily sensations can help us be more present, and explain how the technique improves interoception in this way. Though there seem to be no scientific studies as of yet specifically tying the Alexander method to improved interoception, it's a likely candidate.

ROSEN METHOD BODYWORK

Rosen Method Bodywork (RMB), developed by Marion Rosen, is a form of touch–talk therapy meant to enhance one's awareness of present-moment felt experience. It is based on the philosophy that muscle tension and suppressed emotion, which we often employ during times of stress, can be released simply by becoming aware of them. It has been shown to be effective for people with chronic skeletal or smooth muscle pain and tension. Interoception is the process that facilitates this practice.

HAKOMI

The Hakomi Method is an experiential psychotherapy method that facilitates access to "core material" (unconscious emotions, memories, beliefs) through present, felt experience. It involves five steps—create a healing relationship, establish mindfulness, evoke experience, processing, transformation, and integration—and has been scientifically validated by the European Association for Psychotherapy. Interoception is a critical part of this practice.

SENSORY AWARENESS

Later called Somatic Awareness, this therapy was developed in Germany by Elsa Gindler in the 20th century. Gindler encouraged clients to draw on the natural activities of everyday life as material for her classes, helping them "sense their way" by paying attention to their inner experience as often as possible throughout each day. Her work influenced students of Sigmund

Freud and Erich Fromm. Some consider Gindler to be the grandmother of somatic psychotherapy.

SOMATIC EXPERIENCING

Somatic Experiencing is a relatively novel form of trauma therapy, developed in the 1970s, that guides clients' attention to interoceptive, kinesthetic, and proprioceptive experience. It has been tied to interoception in the research literature, underscoring the connection between interoceptive awareness and sense of self. According to its founders, Somatic Experiencing is "designed to direct the attention of the person to internal sensations that facilitate biological completion of thwarted responses, thus leading to resolution of the trauma response and the creation of new interoceptive experiences of agency and mastery."

BREATH THERAPY

Breath therapy, or breath work, has become a popular form of alternative therapy that involves harnessing deliberate control and awareness of the breath to reach altered states of consciousness and/or emotional and psychological well-being. One of the best-known methods is Holotropic Breathwork, developed by Stan Grof, which shares many features with psychedelic-induced altered states of consciousness. As another example, rebirthing breath work is a modality used in the Wayapa wellness practices of Indigenous Australians and involves unraveling the birth–death cycle through conscious connected breathing. Awareness of the breath is an interoceptive process, and many studies on breath work mention interoception, if not explicitly measure it.

SOCIAL JUSTICE APPROACHES

Many movers and shakers within social justice movements have created embodiment practices informed by differences in historical, cultural, racial, ethnic, and gender-based and/or sexual identities. While there is limited research on the relationship between interoception as a scientific construct and these practices, we can safely assume (and must assume, if we are to make interoceptive techniques diverse and inclusive) a strong connection. The following people are pioneers in this field and should be followed closely for a fuller understanding of how interoception might differ for people with marginalized bodies.

SCIENCE OF SOCIAL JUSTICE

Sará King, PhD, is the creator of the Science of Social Justice framework for research and facilitation as well as the Systems-Based Awareness Map, a model of the relationship between individual and collective awareness and well-being. As noted in her professional website bio, she has dedicated much of her career to practicing and developing clinical and field-based research on how contemplative practices grounded in social justice can be supportive to a wide variety of organizations and social environments. Considered an international thought leader in the field of study that merges neuroscience, mindfulness, and social justice research, King specializes in applying this research to help alleviate health disparities, foster innovation, and nurture healing interpersonal relationships for clientele.

"If you have a body, you deserve well-being," she says in a presentation for the Wisdom 2.0 Summit. "In my theorizing, social justice and well-being are one and the same thing."

One of her practices, called Creating a Compassionate Container, invites group participants to bring awareness to their breath, orient themselves by noticing their internal experience of the people around them, ground themselves by imagining a root system below their feet connecting all participants, hum together to create a harmonic interpersonal field, and engage in a compassionate viewing practice, noticing how images of different people on a screen elicit different bodily responses.

RESILIENCE TOOLKIT

Nkem Ndefo, MSN, CNM, RN, is the founder of Lumos Transforms and creator of the Resilience Toolkit, a resource that teaches individuals, communities, and organizations how to recognize their own stress and relaxation cycles in order to envision, create, and implement positive change.

As noted in her bio on the Lumos Transforms website, Ndefo has lent her expertise in the trauma-informed and resilience-oriented (TIRO) approach, embodied anti-oppression, and brought collective liberation to a wide range of projects. She served on the Los Angeles County Trauma-and Resilience-Informed Systems Change Initiative Workgroup and developed a pilot violence-prevention academy for peer support workers as part of Los Angeles County Department of Public Health's Trauma Prevention Initiative. Currently, she is leading an embodied diversity, inclusion, and antiracism initiative for the Los Angeles County Department of Health Services, codirecting Embody Lab's Integrative Somatic Trauma Therapy Certification Program, and directing the Resilience Toolkit Certification Training Program.

One essential part of the Resilience Toolkit is the practice of safety, Ndefo says in an interview with Lynn Fraser of the Kiloby Center for Recovery.

> Safety is the most important thing, for everybody. But people define it differently. It's an important starting point conversation that needs to be had, in all settings: what makes you feel safe? It's very hard to embark on a healing journey or transformation journey if you don't have any idea of where you can return and rest, where you can find safety, when things get rough. The starting point for any journey is "Where is home?" And that may be quite an exploration, that in and of itself may be part of the healing. What you find may be so small, a little thing, and it's not super safe but it's relatively safer than everything else around it, and that's a win. How do we keep returning there, over and over again, so that it strengthens and deepens its roots, and widens until it feels solid, and becomes an anchor I can return to as I venture out? That, across the board, is the one thing that holds true with every group, organization, or individual I've ever worked with.

At the same time, Ndefo says, some of her clients use the tools in her Toolkit to reach a feeling of safety, only to find that when their stress returns they can't maintain it. Instead of berating themselves, Ndefo proposes they ask themselves, "What about the places you live and work? Are you actually responding appropriately to the places you live and work, and you're asking yourself to do something superhuman, to be calm and okay in a situation that's not calm and okay? Let's not pile it on the individual . . . you need to look at the system around them."

OPPRESSION AND THE BODY

Christine Caldwell, PhD, is a registered somatic movement educator and therapist. She is the founder of Somatic Counseling Psychology at Naropa University and founder of the Moving Cycle Institute. In her book *Oppression and the Body: Roots, Resistance, and Resolution*, she explores the body as the main site of oppression in Western society. "In a culture where bodies of people who are brown, black, female, transgender, disabled, fat, or queer are often shamed, sexualized, ignored, and oppressed," the book asks, "what does it mean to live in a marginalized body?" Caldwell shows us how power, privilege, and oppression can confine bodily expression for disparaged individuals.

In one section of the book, "Queering/Querying the Body," Caldwell recommends the following exercise:

> Think about a body norm that you enact on a regular basis. Perhaps you sit with your knees held closely together while riding on public transit, or whiten your teeth. Maybe you wear a bra or a suit, smile at strangers on the street, remove hair from a particular area of your body, or take medication for acne. The next time you enact this norm, notice the (perhaps subtle) bodily sensations, emotions, and images that attend this activity. See if you can suspend interpretation of this somatic data long enough to allow this new information to settle into patterns and possible understandings on its own, without imposing meaning or judgment.

PRACTICES FROM THIS BOOK

To summarize the insights from the interviews included in this book, I'd like to draw out the following practices as potentially useful for interoceptive connection.

Kama muta and peak experiences. Pay attention to your body while experiencing kama muta moments, and try to pursue or create experiences that might trigger them. Discuss your insights with others or write about them in a journal.

Nature. Spend time in nature and notice how your body feels. Can you go beyond "calm" and "relaxed" to a more specific description of your inner experience? How does relating to nature feel different, in your body, from relating to people?

Music. Notice your bodily response to music. You may feel emotional or want to dance. Where in your body do you feel this shift? How does it change your mental state? How do you relate to other people through music, as mediated by your body?

Body oneness. Think about the social activities you may normally engage in—eating, drinking, moving together—from the angle of body oneness and movement synchrony. Being aware of the reason it's so powerful, experiencing it in your body, and sharing that awareness with others can enhance the moment.

Simple sensation. Instead of "trying to be happier," introduce a sensation to your body. As Kelly Mahler recommends, start simple, for example by holding an ice cube in your hand. Move on to sensations that feel more pleasurable, like standing outside on the porch and feeling the wind in your hair. Make a list of sensations like these and explore them for their own sake.

Negativity. If, as Kris Oldroyd suggests, our caregivers' responses to negative emotion have so much bearing on our interoceptive awareness, we might do well to pay attention to our own relationship to negative emotion. Pay attention to how it feels and how you tend to manage it, both in yourself and others. Decide for yourself how you want to relate to it and take steps, such as adopting a mindfulness practice, to change your habits.

Compassion. The moment you start to judge someone, as Karen Dobkins says, ask yourself, "Do I do this as well?" and see if you can then approach conflict resolution through connection. Once you relate the behavior to yourself, see if you feel compassion in your body and use that shift to guide your embodied interaction with them.

BODY TRUSTING TOOLKIT

Finally, I'd like to include a few practices that are not scientifically or clinically validated but that I've developed myself for my own interoceptive connection purposes. As this process is already so idiosyncratic and individual, and we're really at the earliest stages of identifying patterns among multiple people, anecdotal personal research may have a great deal of value right now. Here's what has worked for me, both in managing substance dependence and anxiety and in living a healthier life independently of any condition. Please do not take these practices as medical advice, but rather an invitation to explore the possibility space.

Create a buffer. Not everyone will feel safe spending more time in their body, at least at the start. Creating a buffer means first learning to "disidentify" from one's sensations, as Jonathan Gibson put it, to notice them with some distance so that you can just observe them without letting them become your whole reality. The eventual goal is to be able to zoom in on your sensations while zooming out to see why you tend to judge them as good or bad. Importantly, disidentifying is not the same as dissociating: the former is typically within one's control; the latter typically not. Creating a buffer can be achieved through certain body-based mindfulness practices such as Open Monitoring, described above.

Be receptive. This is the secret ingredient, in my opinion. Receptivity is the ultimate antidote to predictive coding, as it's a bottom-up (sensory) orientation, rather than a top-down (cognitive) one. To maximize the possibility space, you have to start from square one. Just feel, without categorizing. Let go of labels. Let go of language. Build a relationship with your body simply by attending to it. Practice paying attention even if you feel "nothing." One reason this step is so important in the beginning is that you have to pay attention to the nothingness in order to respect and trust your body in full; otherwise, you only attend to feelings reactively, like medical systems that only offer healthcare when something goes wrong. Pay attention all the time. To set this step in motion, try Reiki, which encourages nonjudgmental observation of internal sensations (as opposed to thoughts).

When you're ready, bring this attentiveness into your surrounding environment, including your social environment.

Appraise. Once you've created an open, nonjudgmental space, you can start to link sensations to emotions and mental states. Is this particular energetic pattern associated with anger? Loneliness? Ask yourself not "How do I feel when I'm anxious?" but "What is this feeling? Is it anxiety, or something else?" Tell yourself not "This is a horrible feeling I have to get rid of" but "This is a normal, potentially useful feeling and I thank my body for alerting me." What do you feel in your body, when, and why? Put the "why" aside for now if it's too complex. Just notice, and take it all seriously. Don't dismiss any of it, even if you end up deeming a feeling unworthy of acting upon. Log all of it.

Separate reaction from reality. There is a subtle moment—nearly as hard to identify as the moment night becomes day—before a feeling becomes a full-blown emotional state. If you can identify it, you have much more power over how your affective reaction to the world impacts your behavior. Imagine receiving an ambiguous text message from a friend that seems to suggest they're upset with you. You have a reaction: frustration. Instead of letting that reaction become your *reality*, the state of things (read: the friend is upset with you), what if you just observed it, and didn't let it go beyond a reaction? Maybe you misread the message. Wouldn't you like to clear up your confusion before you take action based on your reaction? It's essential to be able to make space between reaction and reality, because you will change your behavior (and thereby your reality) based on what you believe to be the state of things. Your feelings do not always reflect what's true of *the situation*, despite the fact that your feelings are always valid *for you*. Emotions influence behavior; hence, the etymology of the word *emotion*, which means "to move." Catch the feeling and hold onto it as one event within the greater possibility space of your body. You can decide if you want to be moved.

Don't always listen. Noticing all feelings is a worthy goal. But some feelings you'll want to act on and some you'll want to ignore. How to know the difference? Only you will know, but it's worth pointing out that "trusting your body" does not mean acting on its signals all the time. It just means trusting that your body is on your side, and that even unpleasant sensations are there to help you, though they may now be unhelpful conditioned responses you choose to ignore. Once you start linking the "why" to the "what" and "when," you can start recognizing when it's a good idea to act and when it's not. This requires paying close attention during a variety of situations over time.

Trust your body *in the moment*. Trust your body to navigate a stressful event and bring you back to homeostasis in a moment-by-moment fashion. You don't need to prepare so much, or stress so much beforehand, if you know you can use your body as a guide during the event itself. How do you do that? Plan to attend to your signals as they come, and when you feel something unpleasant such as anxiety, be kind and curious. I can't emphasize enough how transformative this has been for me. I certainly still have stressful events now and then that are unpleasant, and I can't seem to prevent that. But 80 percent of the time, I am calm in situations where I used to be anxious. One big reason for this is that I always plan to use my body as a guide, to attend to my bodily sensations in each moment, rather than worry ahead of time whether I'll feel something unpleasant or not. I expect to have a reaction and I'm curious what it will be. It's just information; I'm picking up on patterns. There's never a dull moment when there's always something to read.

Think in terms of homeostasis. Whenever we're distressed, it's our body's way of asking us to come home, to return to homeostasis. Even if the body is home, however, environmental conditions can make it challenging to get there. As Nkem Ndefo says, be gentle with yourself and resist putting the entire weight of regulation onto yourself. What is it in your immediate surroundings that needs to change? It could be your family system or working environment or friendship group or systemic circumstances that needs to change, not you. Seek outside resources, such as Ndefo's Resilience Toolkit, to help you in this process of returning to yourself.

Wait until something feels right. I've been practicing this one more and more since my interview with Jonathan Gibson. If you can afford to, don't make a decision until there's a kind of "open" feeling in your body, as he put it. Sometimes months go by for me without that open feeling, but I recognize it as soon as it's there. For me it feels like a subtle urge to grow toward the decision, like a plant gently moving toward the light. The best things in life have come to me when I've followed this feeling. The more you practice it, the easier it becomes to act on what does or doesn't feel right in any given situation.

Differentiate feeling types. Learn to recognize different shades of the same feeling, e.g., joy or stress, and pay attention to the nuances in how they feel and what events they're connected to rather than categorizing them too quickly as "joy" or "stress" and moving on. Once you start paying attention and reflecting more, you might recognize feelings that you can become curious about, rather than acting on them immediately. For example, it is possible to feel shame, recognize that it is unwarranted, and decide to let it pass rather than allowing it to move you. Still, you honor the fact that you feel that way, and validate yourself. You feel that way. You just do. And it means something. But it may be the case that you don't want to put so much stock in some of your feelings compared to others. And learning to differentiate these feeling types is essential too. In my case, I have counted several different types of anxiety linked to different causes, which happen in different parts of my body. For example, anxiety over the distant past feels like a dull weight in my heart. Anxiety over a person I don't trust feels like a knot around my heart. And premonition anxiety over something that will happen later the same day feels like a little ghost hanging in the pit of my stomach. To notice these anxiety types doesn't mean I experience anxiety more, but rather less, maybe because the emotion feels less like an unwelcome guest and more like someone to get to know better.

Give your feelings some perspective. We tend to examine something more closely, stare at it through a microscope, in order to understand it better. But the more I practice zeroing in on my feelings, the more I realize that distance from them is what clarifies their meaning. It reminds me of Gladwell's remark about drunkenness and myopia: being wrapped up in an immediate feeling, and placing so much importance on it, isn't so different from the short-sightedness of inebriation. Strong feelings I had in childhood, or the experience of loving my first boyfriend, are more important than some vague sense of anxiety or ratio of joy to dissatisfaction I happen to feel today. Give your feelings time, perspective, and kindness.

Glossary

Alexithymia: *lack of emotional awareness or, more specifically, difficulty in identifying and/or describing feelings and in distinguishing physical feelings from the bodily sensations of emotional arousal.*

Attachment: *your style of relating to others based on the emotional bond that formed between you and your caregiver when you were an infant.*

Attention regulation: *1) the ability to self-monitor one's deployment of attention, which includes maintaining attention, ignoring distracting or irrelevant stimuli, staying alert to task goals, and coordinating one's attention during a task; 2) the basis of all meditative and mindfulness techniques.*

Bodily self: *one's sense of self, including self-consciousness and self-awareness, formed by the integration of multisensory signals from the body.*

Body listening: *active listening to the body for insight.*

Body oneness: *the sense of merging one's body with another's, as during synchronous movement or sharing food and drink.*

Body trusting: *experiencing one's body as safe and trustworthy.*

Connectedness: *the sense of feeling connected to oneself, others, or the cosmos.*

Critical period: *a developmental stage in early life when the nervous system is especially sensitive to certain environmental stimuli.*

Emotional awareness: *your ability to identify which emotions you're experiencing at any given time.*

Engram: *a physical memory trace in the brain and/or body.*

Entactogen: *a drug that produces a "touching within," meaning an enhanced ability to access inner feeling states.*

Flexible switching: *a dynamic social exchange during which each person regularly checks in with their feeling state in response to the other's words, body language, energy, etc.*

Homeostasis: *the ability to maintain a relatively stable internal state that persists despite changes in the world outside.*

Insular cortex: *a part of the brain involved in multisensory integration, sense of self, social cognition, interoception, and emotion regulation.*

Interoception: *the process of sensing the body from within.*

Interoceptive accuracy: *an objective measure of how accurate you are at detecting and identifying bodily signals, such as heartbeat perception.*

Interoceptive awareness: *a metacognitive measure of how aware you are of your own interoceptive accuracy.*

Interoceptive sensibility: *how you think about what you feel; your "interoceptive style."*

Interpersonal emotion regulation: *the process of navigating the emotional experience of oneself or another person through social interaction.*

Kama muta: *the feeling of being moved by love, often accompanied by a warm heart, lump in throat, tears in eyes, goosebumps, or some combination of these.*

Movement synchrony: *synchronized movement of two or more bodies.*

Not-distracting: *the tendency not to ignore or distract oneself from sensations of pain or discomfort.*

Noticing: *awareness of uncomfortable, comfortable, and neutral body sensations.*

Not-worrying: *tendency not to worry or experience emotional distress with sensations of pain or discomfort.*

Open state: *a state, often produced under MDMA and classic psychedelics, whereby one's defenses are lowered and deeper emotion states more easily accessed.*

Peak experience: *an altered state of consciousness characterized by euphoria; a transcendent moment of joy or elation.*

Possibility space: *what's not predicted by the body or brain; alternative feelings and thoughts that deviate from one's typical experience.*

Predictive coding: *the theory that the brain is constantly generating and updating a mental (and physical) model of the environment; a model for human brain function that places prediction, anticipation, and expectation at the helm of neural processing.*

Prediction error: *the failure of an expected event to occur; especially, the brain's observation of said failure.*

Reappraisal: *reinterpreting the meaning of a feeling to change its emotional trajectory.*

Relational health: *the quality of one's relationship to one's body, oneself, other people, and/or the environment.*

Self–other distinction: *your cognitive and interoceptive sense of being a distinct social actor in the world, separate from others.*

Self-regulation: *your ability to monitor and manage your energy states, emotions, thoughts, and behaviors.*

Social interoception: *the process of sensing the body from within and using or communicating what's experienced during a social encounter.*

Notes

Introduction

he writes in his 2017 paper "The Pain of Granting Otherness: Interoception and the Differentiation of the Object" Joona Taipale, "The Pain of Granting Otherness: Interoception and the Differentiation of the Object," *Gestalt Theory* 39 (2017), https://doi.org/10.1515/gth-2017-0013.

a fundamental property of social contact throughout the life span would be to enhance self-awareness Nesrine Hazem et al., "Social Contact Enhances Bodily Self-Awareness," *Scientific Reports* 8, 4195 (2018), https://doi.org/10.1038/s41598-018-22497-1.

One of the main drivers behind the therapeutic effects of psychedelics is now thought to be a change in one's experience of social relationships Leor Roseman et al., "Editorial: Psychedelic Sociality: Pharmacological and Extrapharmacological Perspectives," *Frontiers in Pharmacology* 13:979764 (2022), https://doi.org/10.3389/fphar.2022.979764.

a region of the brain thought to drive self-referential thinking Saga Briggs, "Dissolving Ego Dissolution: Rethinking the Role of the Default Mode Network in Psychedelics," MIND Foundation, May 14, 2021, https://mind-foundation.org/ego-dissolution/.

in an interview with humanistic psychologist Scott Barry Kaufman David Yaden and Scott Barry Kaufman, "David Yaden on the Science of Self-Transcendent Experiences," April 14, 2020, *The Psychology Podcast*, https://scottbarrykaufman.com/podcast/the-science-of-self-transcendent-experiences-with-david-yaden/.

everything and everyone needs to be approached with love, including myself Michael Pollan, *How to Change Your Mind: What the New Science of Psychedelics Teaches Us About Consciousness, Dying, Addiction, Depression, and Transcendence* (New York: Penguin, 2018).

the relationship between psilocybin and connectedness in participants with depression Rosalind Watts et al., "The Watts Connectedness Scale: A New Scale for Measuring a Sense of Connectedness to Self, Others, and World," *Psychopharmacology* 239, no. 11 (2022): 3461–83, https://doi.org/10.1007/s00213-022-06187-5.

Part I: Predicting the Body

too much prediction in the wrong direction can trap us into destructive belief systems and habits, including substance abuse Mark Miller, Julian Kiverstein, and Erik Rietveld, "Embodying Addiction: A Predictive Processing Account," *Brain and Cognition* 138 (2020): 105495, https://doi.org/10.1016/j.bandc.2019.105495.

Lisa Feldman Barrett, PhD, who pioneered the EPIC (Embodied Predictive Interoception Coding) model of cognition Lisa Feldman Barrett and W. Kyle Simmons, "Interoceptive Predictions in the Brain," *Nature Reviews Neuroscience* 16,7 (2015): 419–29, https://doi.org/10.1038/nrn3950.

eliminating nuance and preserving the gist of the experience Jordana Cepelewicz, "To Make Sense of the Present, Brains May Predict the Future," *Quanta Magazine*, July 10, 2018, https://www.quantamagazine.org/to-make-sense-of-the-present-brains-may-predict-the-future-20180710/.

neuroscientists Karen Dobkins, PhD, and Andy Arnold, PhD, write in a theoretical paper on the topic Karen Dobkins and Andy Arnold, "Social Interoception: Listening to Your Body Is Key for Social Connection" (unpublished manuscript, February 2020), e-mail.

"It was actually PTSD that made me first get interested in the heart and interoception," she says Ricardo Lopes and Sarah Garfinkel, "Sarah Garfinkel: Interoception, Emotion, and Mental Health," *The Dissenter*, April 30, 2020, https://youtu.be/ZigEDvbsGb8.

the child will find ways to avoid feeling them, and develop a distorted sense of interoception Saga Briggs, "Social Interoception: The Case for Treating Mental Illnesses Through the Body, in a Social Setting," MIND Foundation, May 1, 2020, https://mind-foundation.org/social-interoception/.

highly correlated with how connected their children were to their bodies Kristina Oldroyd, Monisha Pasupathi, and Cecilia Wainryb, "Social Antecedents to the Development of Interoception: Attachment Related Processes Are Associated with Interoception," *Frontiers in Psychology* 10: 712 (2019), https://doi.org/10.3389/fpsyg.2019.00712.

controlling for child gender and ethnicity, family income, maternal stress, and the above maternal socialization factors Jennifer K. MacCormack et al., "Mothers' Interoceptive Knowledge Predicts Children's Emotion Regulation and Social

Skills in Middle Childhood," *Social Development* 29 (2020): 578–99, https://doi.org/10.1111/sode.12418.

distinct from a person's subjective feelings about the body Sarah N. Garfinkel and Hugo D. Critchley, "Interoception, emotion and brain: new insights link internal physiology to social behaviour. *Commentary on::* 'Anterior insular cortex mediates bodily sensibility and social anxiety' by Terasawa et al. (2012)," *Social cognitive and affective neuroscience* vol. 8,3 (2013): 231-4, https://doi.org/10.1093/scan/nss140.

Interoceptive sensibility (IS) refers to an individual's style of interpreting their bodily sensations Sarah N. Garfinkel et al., "Knowing Your Own Heart: Distinguishing Interoceptive Accuracy from Interoceptive Awareness," *Biological Psychology,* 104 (January 2015), 65–74, https://doi.org/10.1016/j.biopsycho.2014.11.004; Thomas Forkmann et al., "Making Sense of What You Sense: Disentangling Interoceptive Awareness, Sensibility and Accuracy," *International Journal of Psychophysiology,* vol. 109 (2016): 71–80, https://doi.org/10.1016/j.ijpsycho.2016.09.019

a metacognitive measure that quantifies individuals' explicit knowledge of and confidence in their interoceptive accuracy Ibid.

lower insular volume and smaller surface area than control groups Lena Lim, Joaquim Radua, MD, and Katya Rubia, PhD, "Gray Matter Abnormalities in Childhood Maltreatment: A Voxel-Wise Meta-Analysis." *The American Journal of Psychiatry* Vol. 171,8 (2014): 854–63, https://doi.org/10.1176/appi.ajp.2014.13101427; Julia M. Sheffield et al., "Reduced Gray Matter Volume in Psychotic Disorder Patients with a History of Childhood Sexual Abuse," *Schizophrenia Research* Vol. 143,1 (2013): 185–91, https://doi.org/10.1016/j.schres.2012.10.032; Simone Kühn and Jürgen Gallinat, "Gray Matter Correlates of Posttraumatic Stress Disorder: A Quantitative Meta-Analysis." *Biological Psychiatry* vol. 73,1 (2013): 70–4, https://doi.org/10.1016/j.biopsych.2012.06.029.

decreased insular activation in response to stimuli than do securely attached individuals C. Nathan DeWall et al., "Do Neural Responses to Rejection Depend on Attachment Style? An fMRI Study," *Social Cognitive and Affective Neuroscience*, Volume 7, Issue 2, February 2012, Pages 184–192, https://doi.org/10.1093/scan/nsq107.

the anterior cingulate cortex (ACC) has been shown to fail to fully integrate with the insula Delia Lenzi et al, "Neural Basis of Attachment-Caregiving Systems Interaction: Insights from Neuroimaging Studies," *Frontiers in Psychology*, Volume 6 (2015):1241, https://doi.org/10.3389/fpsyg.2015.01241.

correlated with a blunted emotional affect and hypo-activating strategies in the face of distress Hugo D. Critchley et al, "Neural Systems Supporting Interoceptive Awareness," *Nature Neuroscience* 7, 189–195 (2004), https://doi.org/10.1038/nn1176.

heightened sensitivity to bodily cues was at the root of anxiety Martin P. Paulus and Murray B. Stein, "Interoception in Anxiety and Depression," *Brain Structure and Function*, 214, 451–463 (2010). https://doi.org/10.1007/s00429-010-0258-9.

that anxiety symptoms arise from discrepancies between a person's actual and expected bodily state Ibid.

being arrested by the incredibly vibrant detail of a daffodil rather than passing it off as just another flower in the garden Sarit Pink-Hashkes, Iris van Rooij, and Johan Kwisthout, "Perception Is in the Details: A Predictive Coding Account of the Psychedelic Phenomenon," *Proceedings of the 39th Annual Meeting of the Cognitive Science Society* (London, July 26–29), pp. 2907–12. London: Cognitive Science Society, 2017.

Psychedelic drugs perturb universal brain processes that normally serve to constrain neural systems central to perception, emotion, cognition, and sense of self Link R. Swanson, "Unifying Theories of Psychedelic Drug Effects," *Frontiers in Pharmacology* 9: 172 (March 2, 2018), https://doi.org/10.3389/fphar.2018.00172.

so they bump people out of their depression/anxiety rut Philip Corlett, e-mail message to author discussing Benjamin Kelmendi et al., "The Role of Psychedelics in Palliative Care Reconsidered: A Case for Psilocybin," *Journal of Psychopharmacology* vol. 30,12 (2016): 1212–1214, https://doi.org/10.1177/0269881116675781.

writes a team at the University of Zurich in Switzerland Jasmine T. Ho, Katrin H. Preller, and Bigna Lenggenhager, "Neuropharmacological Modulation of the Aberrant Bodily Self through Psychedelics," *Neuroscience and Biobehavioral Reviews* 108 (2020): 526–41, https://doi.org/10.1016/j.neubiorev.2019.12.006.

A neurocognitive mechanism which may underlie the effects of psilocybin on emotion processing and interoceptive awareness can be found in the predictive processing framework Josephine Marschall et al., "Psilocybin Microdosing Does Not Affect Emotion-Related Symptoms and Processing: A Preregistered Field and Lab-Based Study," *Journal of Psychopharmacology* 36: 1 (2022); 97–113, https://doi.org/10.1177/0269881121105.

Although a region of the brain called the thalamus is thought to mediate this influx of information Franz X. Vollenweider and Mark A. Geyer, "A Systems Model of Altered Consciousness: Integrating Natural and Drug-Induced Psychoses," *Brain Research Bulletin* 56(2001): 495–507, https://doi.org/10.1016/S0361-9230(01)00646-3.

In a 2017 MAPS study, following research correlating heightened insula activation with PTSD and Social Anxiety Ishan C. Walpola et al., "Altered Insula Connectivity under MDMA," *Neuropsychopharmacology* 42 (2017): 2152–62, https://doi.org/10.1038/npp.2017.35.

Part II: Disconnection

implicated in just about every affliction under the sun, including substance abuse, psychosomatic disorders, anxiety, depression, eating disorders, addiction, OCD, bipolar disorder, schizophrenia, and the distress surrounding infertility, Claire Lamas et al., "Alexithymia in Infertile Women," *Journal of Psychosomatic Obstetrics and Gynecology* 27:1 (2006): 23–30, https://doi.org/10.1080/01674820500238112

There is even one study linking alexithymia to cell phone addiction Songli Mei et al., "The Relationship between College Students' Alexithymia and Mobile Phone Addiction: Testing Mediation and Moderation Effects," *BMC Psychiatry* 18: 329 (2018), https://doi.org/10.1186/s12888-018-1891-8.

higher malleability of body representation in illusions of body-ownership Sophie Betka et al., "How Do Self-Assessment of Alexithymia and Sensitivity to Bodily Sensations Relate to Alcohol Consumption?" Alcohol Clinical and Experimental Research, 42: 81–88, November 2, 2017, https://doi.org/10.1111/acer.13542

In the 2008 paper "How Do Self-Assessment of Alexithymia and Sensitivity to Bodily Sensations Relate to Alcohol Consumption?" Ibid.

consuming nearly twice as much caffeine per day as the other students Michael Lyvers, Natalija Duric, and Fred Arne Thorberg, "Caffeine Use and Alexithymia in University Students," *Journal of Psychoactive Drugs* 46: 4, (2014): 340–46, https://doi.org/10.1080/02791072.2014.942043.

Parenting style, notably poor maternal care, and avoidant attachment, predict the later expression of alexithymia across patient groups Cecilia Serena Pace et al., "When Parenting Fails: Alexithymia and Attachment States of Mind in Mothers of Female Patients with Eating Disorders," *Frontiers in Psychology* 6 (2015), https://doi.org/10.3389/fpsyg.2015.01145.

it's thought to be a very, very valuable adjunct to psychotherapy "The Science and Shamanism of Psychedelics," *Goop*, January 10, 2019, https://goop.com/wellness/health/the-science-and-shamanism-of-psychedelics/.

I think there's something here and this should be looked into much more Shayla Love, "People Born Blind Are Mysteriously Protected from Schizophrenia," *Vice*, February 11, 2020, https://www.vice.com/en/article/939qbz/people-born-blind-are-mysteriously-protected-from-schizophrenia.

People high in alexithymia are more susceptible to the illusion, struggling to integrate simultaneous sensory events into a single experience Delphine Grynberg and Olga Pollatos, "Alexithymia Modulates the Experience of the Rubber Hand Illusion," *Frontiers in Human Neuroscience* 9: 357 (June 18, 2015), https://doi.org/10.3389/fnhum.2015.00357.

alexithymia is associated with an abnormal focus of one's own body Ibid.

"The rubber hand illusion is quantitatively and qualitatively stronger in schizophrenia," writes a team from Vanderbilt University's Department of Psychology Katharine N. Thakkar et al., "Disturbances in Body Ownership in Schizophrenia: Evidence from the Rubber Hand Illusion and Case Study of a Spontaneous Out-of-Body Experience," *PloS One* 6(10): e27089, October 31, 2011, https://doi.org/10.1371/journal.pone.0027089.

Another group of researchers, from Italy, found that during the illusion Francesco della Gatta et al., "Decreased Motor Cortex Excitability Mirrors Own Hand Disembodiment During the Rubber Hand Illusion," *eLife* 5:e14972, October 20, 2016, https://doi.org/10.7554/eLife.14972.

"People with schizophrenia typically do far better in poorer nations such as India, Nigeria, and Colombia than in Denmark, England, and the United States" Shankar Vedantam, "Social Network's Healing Power Is Borne Out in Poorer Nations," *Washington Post*, June 27, 2005, https://www.washingtonpost.com/archive/politics/2005/06/27/social-networks-healing-power-is-borne-out-in-poorer-nations/244d1ffe-3765-4a28-82e4-c0b847dbccc6/.

where most patients are homeless, in group homes or on their own, in psychiatric facilities or in jail Ibid.

found that medications "primarily controlled patients' delusions and hallucinations, not the 'negative' symptoms that cause patients to disappear into silent, inner worlds" William T. Carpenter Jr., John S. Strauss, and John J. Bartko, "Flexible System for the Diagnosis of Schizophrenia: Report from the WHO International Pilot Study of Schizophrenia," *Science*, Vol. 182, 4118: 1275–8, December 21, 1973 https://doi.org/10.1126/science.182.4118.1275

people with low resilience to stress "show reduced attention to bodily signals but greater neural processing to aversive bodily perturbations" Lori Haase et al., "When the Brain Does Not Adequately Feel the Body: Links Between Low Resilience and Interoception," *Biological Psychology* 113: 37–45, January 2016, https://doi.org/10.1016/j.biopsycho.2015.11.004.

neurobiologists at Thomas Jefferson University in Philadelphia found that too much isolation changes the structure of the brain V. Heng, M. J. Zigmond, R. J. Smeyne, "Neurological Effects of Moving from an Enriched Environment to Social Isolation in Adult Mice" (Session 291.02/K14. 2018 Neuroscience Meeting Planner, San Diego, CA: Society for Neuroscience, November 5, 2018, https://www.abstractsonline.com/pp8/#!/4649/presentation/20940).

in a 2018 panel discussion at the annual Society for Neuroscience conference Dana G. Smith, "Neuroscientists Make a Case against Solitary Confinement," *Scientific American*, November 9, 2018, https://www.scientificamerican.com/article/neuroscientists-make-a-case-against-solitary-confinement/.

accounting for other related factors such as self-esteem, gratitude, subjective well-being, depression, and alexithymia Andy J. Arnold and Karen Dobkins, "Trust Some Body: Loneliness Is Associated with Altered Interoceptive Sensibility," https://quote.ucsd.edu/kdobkins/files/2019/05/Arnold-Dobkins-2019.pdf.

"by incorporating an 'embodied' experience of drug use together with the individual's predicted versus actual internal state to modulate approach or avoidance behavior, i.e. whether to take or not to take drugs" Martin P. Paulus and Jennifer L. Stewart, "Interoception and Drug Addiction," *Neuropharmacology* 76, Pt B,0 0 (2014): 342–50, January 2014, https://doi.org/10.1016/j.neuropharm.2013.07.002.

"These results point to the importance of reinstatement of social reward in the treatment of stimulant addiction" Katrin H. Preller et al., "Functional Changes of the Reward System Underlie Blunted Response to Social Gaze in Cocaine Users," *Proceedings of the National Academy of Sciences of the United States of America* 111, 7: 2842–7, January 21, 2014, https://doi.org/10.1073/pnas.1317090111.

"identified social factors, i.e., smoking as a way of connecting with other people, that contributed to their addiction" Matthew W. Johnson, Albert Garcia-Romeu, and Roland R. Griffiths, "Long-Term Follow-up of Psilocybin-Facilitated Smoking Cessation," *American Journal of Drug and Alcohol Abuse* 43: 1: 55–60, January 2017, https://doi.org/10.3109/00952990.2016.1170135.

study has also shown that MDMA is effective in treating alcoholism Ben Sessa et al., "First Study of Safety and Tolerability of 3,4-methylenedioxymethamphetamine-assisted Psychotherapy in Patients with Alcohol Use Disorder," *Journal of Psychopharmacology* 35(4): 375–83, April 2021, https://doi.org/10.1177/0269881121991792.

In a Medium post about alcohol and stress Markham Heid, "What Alcohol Does to a Stressed-Out Brain," Elemental, Medium, April 30, 2020, https://elemental.medium.com/what-alcohol-does-to-a-stressed-out-brain-af3771a36dba.

The brain's normal non-alcohol state begins to change, and in some cases it may become a more anxious one Ibid.

the persistent triggering of abnormally large somatic errors by sensory signals (either real or imagined) irrespective of context—that is, context rigidity Paulus and Stein, "Interoception in Anxiety and Depression."

Metabolic studies and post-mortem studies have also found differences in the insular cortex, for instance, in the expression of neurotransmitters released by neurons in this part of the brain Richard Gray, "'Island of the Brain' Explains How Physical States Affect Anxiety," *Horizon*, Health: Interview, August 2, 2018, https://ec.europa.eu/research-and-innovation/en/horizon-magazine/island-brain-explains-how-physical-states-affect-anxiety.

The New York Times *reported recently that even "light activity—walking at a casual pace, shopping, playing an instrument, doing chores around the house—has a big effect [on mental health]"* Perri Klass, MD, "The Benefits of Exercise for Children's Mental Health," *New York Times*, March 2, 2020, https://www.nytimes.com/2020/03/02/well/family/the-benefits-of-exercise-for-childrens-mental-health.html.

Movement is intimately connected to the brain's creation of a sense of body ownership Luke Miller and Alessandro Farnè, "Cognition: Losing Self Control," *eLife* vol. 5 e21404, October 20, 2016, https://doi.org/10.7554/eLife.21404.

At the same time, "impairment of the motor system directly affects the multisensory sense of body ownership." Elena Nava et al., "Action Shapes the Sense of Body Ownership Across Human Development," *Frontiers in Psychology* vol 9, December 17, 2018, https://doi.org/10.3389/fpsyg.2018.02507.

the central role of the attentional control of bodily awareness, and awareness of breathing in particular, during various contemplative practices Raphaël Millière et al., "Psychedelics, Meditation, and Self-Consciousness," *Frontiers in Psychology* vol. 9 1475. 4 September 4, 2018, https://doi.org/10.3389/fpsyg.2018.01475.

a high-level multisensory hub engaged in body and action awareness in general Silvia Seghezzi, Gianluigi Giannini, and Laura Zaparoli, "Neurofunctional Correlates of Body-Ownership and Sense of Agency: A Meta-Analytical Account of self-Consciousness," *Cortex* vol. 121, December 2019: 169–178, https://doi.org/10.1016/j.cortex.2019.08.018.

Michael Jawer in Psychology Today *writes* Michael Jawer, "PTSD: A Window into the Bodymind (Part 2)," *Psychology Today*, February 18, 2013, https://www.psychologytoday.com/us/blog/feeling-too-much/201302/ptsd-window-the-bodymind-part-2.

researchers found that "interoceptive brain capacity enhanced by Mindfulness Based Stress Reduction appears to be the primary cerebral mechanism that regulates emotional disturbances and improves anxiety symptoms of PTSD" Seung Suk Kang, Scott R. Sponheim, and Kelvin O. Lim, "Interoception Underlies Therapeutic Effects of Mindfulness Meditation for Post-Traumatic Stress Disorder: A Randomized Clinical Trial," *Biological Psychiatry: Cognitive Neuroscience and Neuroimaging* 7, August 2022, 7(8):793–804, https://doi.org/10.1016/j.bpsc.2021.10.005.

"It is suggested that higher levels of PTSD affect the ability of veterans to initiate and maintain interpersonal relationships and that these interpersonal problems are evident in poorer levels of family functioning and poorer dyadic adjustment" Carol MacDonald et al., "Posttraumatic Stress Disorder and Interpersonal Functioning in Vietnam War Veterans: A Mediational Model," *Journal of Traumatic Stress* vol, 12, issue 4, October 1999: 701–707, https://doi.org/10.1023/A:1024729520686.

clinicians Peter Gasser, MD, and Peter Oehen, MD, in a 2022 paper examining the effects of MDMA and LSD group therapy on c-PTSD Peter Oehen and Peter Gasser, "Using a MDMA- and LSD-Group Therapy Model in Clinical Practice in Switzerland and Highlighting the Treatment of Trauma-Related Disorders," *Frontiers in Psychiatry* 13: 863552 (April 25, 2022), https://doi.org/10.3389/fpsyt.2022.863552.

(i.e., eating restrictions) that amplify autonomic hunger signals to minimize interoceptive uncertainty and maintain a more coherent sense of (interoceptive) self. Laura Barca and Giovanni Pezzulo, "Keep Your Interoceptive Streams under Control: An Active Inference Perspective on Anorexia Nervosa," *Cognitive, Affective, & Behavioral Neuroscience* 20: 2 (February 7, 2020): 427–40. https://doi.org/10.3758/s13415-020-00777-6.

Interoceptive reliance, like the ability to trust, positively consider, and positively use inner sensations, should be a privileged target of psychotherapeutic interventions in obesity Clémence Willem et al., "Interoceptive Reliance as a Major Determinant of Emotional Eating in Adult Obesity," *Journal of Health Psychology* 26: 12 (January 31, 2021): 2118–30, https://doi.org/10.1177/1359105320903093.

"not feeling safe in one's body" as the central bridge between interoception and eating disorder symptoms Tiffany A. Brown et al., "Body Mistrust Bridges Interoceptive Awareness and Eating Disorder Symptoms," *Journal of Abnormal Psychology* 129: 5 (2020): 445–56, https://doi.org/10.1037/abn0000516.

they are also characterized by divergent traits (e.g., avoidance versus approach; inhibited versus impulsive) that might suggest opposite patterns of response to the anticipation and experience of interoceptive stimuli Christina E. Wierenga et al., "Increased Anticipatory Brain Response to Pleasant Touch in Women Remitted from Bulimia Nervosa," *Translational Psychiatry* 10: 236 (July 16, 2020), https://doi.org/10.1038/s41398-020-00916-0.

the somatic dimensions of other conditions such as depression and anxiety Laura Crucianelli et al., "The Effect of Intranasal Oxytocin on the Perception of Affective Touch and Multisensory Integration in Anorexia Nervosa: Protocol for a Double-Blind Placebo-Controlled Crossover Study," *BMJ Open*, 9:e024913 (March 15, 2019), http://dx.doi.org/10.1136/bmjopen-2018-024913.

body trust has been shown to dip temporarily during adolescence Jennifer Todd et al., "An Exploration of the Associations Between Facets of Interoceptive Awareness and Body Image in Adolescents," *Body Image* vol. 31 (December 2019): 171–180, https://doi.org/10.1016/j.bodyim.2019.10.004.

found that AN is associated with difficulties inferring others' emotional states "despite largely intact nonemotional mental state inference" Timo Brockmeyer et al., "Social Cognition in Anorexia Nervosa: Specific Difficulties in Decoding Emotional but Not Nonemotional Mental States," *International Journal of Eating Disorders* 49:9 (September 2016): 883–90, https://doi.org/10.1002/eat.22574.

as mental health journalist Marianne Apostolides writes, "efforts to provide rational reasons for its development seem incapable of changing the behavior" Marianne Apostolides, "Psychedelics Offer New Route to Recovery from Eating Disorders," *Proto.Life*. February 3, 2022, https://proto.life/2022/02/psychedelics-offer-new-route-to-recovery-from-eating-disorders/.

They report less interoceptive sensibility, suggesting that they use this information less, in terms of a reduced ability to regulate body-related attention or use body sensations for distress regulation Thomas Forkmann et al., "Sense It and Use It: Interoceptive Accuracy and Sensibility in Suicide Ideators," *BMC Psychiatry* 19: 334 (November 1, 2019), https://doi.org/10.1186/s12888-019-2322-1.

the first research institution to record electrical activity from the von Economo neurons (or "empathy neurons") in live human tissue Rachel Tompa, "New Clues About a Huge, Rare, Human Brain Cell," Allen Institute, Brain Science, March 3, 2020, https://alleninstitute.org/what-we-do/brain-science/news-press/articles/new-clues-about-huge-rare-human-brain-cell.

"They're big neurons, which I think do a very fast read of something and then relay that information elsewhere quickly" Ingfei Chen, "Brain Cells for Socializing," *Smithsonian Magazine*, June 2009, https://www.smithsonianmag.com/science-nature/brain-cells-for-socializing-133855450/.

"Insula perfusion . . . is progressively decreased in alcohol-addicted individuals, and alcoholism is also associated with a loss of insula gray matter Markus Heilig et al., "Time to Connect: Bringing Social Context into Addiction Neuroscience," *Nature Reviews: Neuroscience* 17: 9 (September 2016): 592–9, https://doi.org/10.1038/nrn.2016.67.

they are more plentiful, helping these individuals preserve their memory capacity and stave off cognitive decline Wanda Thibodeaux, "Are You a 'Superager'? You Might Be If Your Brain Shows These 2 Key Traits," Inc., Innovate, February 23, 2018, https://www.inc.com/wanda-thibodeaux/heres-how-brains-of-superagers-are-different-than-minds-of-normal-people.html.

people who take their own lives tend to have a greater number of them, more densely packed Charles Q. Choi, "Neurons Offer Clues to Suicide," *Scientific American*, Mind & Brain, November 1, 2011, https://www.scientificamerican.com/article/suicide-cells/.

Part III: Connection

the conflict between our need for attachment and our need for authenticity Gabor Mate, "Authenticity," Paloma Foundation, recorded November 15, 2012, posted April 9, 2017, YouTube, https://youtu.be/G57xfseSUTU.

borrowed from the ancient Sanskrit where it meant "moved by love" Alan Page Fiske, *Kama Muta: Discovering the Connecting Emotion* (London: Routledge, 2020), https://doi.org/10.4324/9780367220952.

"Connectedness" is not a scientific construct The Watts Connectedness Scale, used in the context of psychedelic therapy, has been developed since this conversation.

uses the terms peak *and* transcendent *experience interchangeably to describe a feeling of unity with everything* Scott Barry Kaufman, *Transcend: The New Science of Self-Actualization* (New York: Tarcher Perigee, 2020).

David Yaden calls them "self-transcendent" experiences, or "those profound moments of connection with something greater than oneself" "Learn," The Varieties Corpus, accessed March 22, 2023, https://www.varietiescorpus.com/learn.

you are automatically benefiting others, and when you are altruistic, you are automatically rewarding and gratifying yourself Abraham H. Maslow and John J. Honigmann, "Synergy: Some Notes of Ruth Benedict," *American Anthropologist* 72: 2 (1970): 320–33, http://www.jstor.org/stable/671574.

changes in the superior parietal lobe, a region of the brain associated with spatial body awareness and social cognition David Bryce Yaden, Jonathan Iwry, and Andrew B. Newberg, "Neuroscience and Religion: Surveying the Field," in *Religion: Mental Religion*, ed. Niki Kasumi Clements (Farmington Hills, MI: Macmillan Reference USA, 2017), 277–299.

"This feeling of unity may result in attributing social qualities to one's spatial environment—a social/spatial conflation" David Bryce Yaden et al., "The Varieties of Self-Transcendent Experience," *Review of General Psychology* vol 21, issue 2 (June 2017), https://doi.org/10.1037/gpr0000102.

"Social-relational emotions, especially kama muta, seem to be salient in experiences of connection with nature." Evi Petersen, Alan Page Fiske, Thomas W. Schubert, "The Role of Social Relational Emotions for Human-Nature Connectedness," *Frontiers in Psychology* 10:2759 (December 2019), https://doi.org/10.3389/fpsyg.2019.02759.

Abraham Maslow's peak experience studies in the 1960s showed that 82 percent of participants had experienced the beauty of nature in a deeply moving way Abraham H. Maslow, *Religions, Values, and Peak-Experiences* (Columbus: Ohio State University Press, 1964).

In 1998, ecopsychologist Jody Davis, PhD, proposed the term transpersonal experiences in nature, which includes the experience of peace, joy, love, support, inspiration, and communion J. Davis, "The Transpersonal Dimensions of Ecopsychology: Nature, Nonduality, and Spiritual Practice," *The Humanistic Psychologist* 26: 1–3 (1998): 69–100, https://doi.org/10.1080/08873267.1998.9976967.

In 2005, Marshall outlined mystical experiences in which people feel that the natural world evokes a sense of unity, knowledge, self-transcendence, eternity, light, and love Paul Marshall, *Mystical Encounters with the Natural World: Experiences and Explanations* (Oxford: Oxford University Press, 2005).

found that contact with nature, emotion, meaning, compassion, and beauty are pathways to improving nature connectedness Ryan Lumber, Miles Richardson, and David Sheffield, "Beyond Knowing Nature: Contact, Emotion, Compassion, Meaning, and Beauty Are Pathways to Nature Connection," *PLoS ONE* 12(5): e0177186 (2017), https://doi.org/10.1371/journal.pone.0177186.

And in 2018, C. L. Anderson found that experiencing gratitude and awe while in nature predicted stress reduction and increases in well-being among military veterans and youth in underserved communities C. L. Anderson, M. Monroy, and D. Keltner, "Awe in Nature Heals: Evidence from Military Veterans, At-Risk Youth, and College Students," *Emotion* 18:8 (2018):1195–1202, https://doi.org/10.1037/emo0000442.

a state of shortsightedness in which superficially understood, immediate aspects of experience have a disproportionate influence on behavior and emotion R. A. Josephs and C. M. Steele, "The Two Faces of Alcohol Myopia: Attentional Mediation of Psychological Stress," *Journal of Abnormal Psychology* vol. 99: 2 (1990), 115–26, https://doi.org/10.1037/0021-843X.99.2.115.

If I have relatively better interoception than some others, the boundary between myself and others is more distinct Stated in reference to the following paper: Clare E. Palmer and Manos Tsakiris, "Going at the Heart of Social Cognition: Is There a Role for Interoception in Self-Other Distinction?" *Current Opinion in Psychology* vol. 24 (December 2018): 21–26, https://doi.org/10.1016/j.copsyc.2018.04.008.

we suggest this behavior is particularly disadvantageous in the realm of social connection Karen Dobkins and Andy Arnold, "Social Interoception: Listening to Your Body Is Key for Social Connection" (unpublished manuscript, February 2020). E-mail.

the two may be **integrated** *for adaptive learning in social situations* Andrew J. Arnold, Piotr Winkielman, and Karen Dobkins, "Interoception and Social Connection," *Frontiers in Psychology* vol. 10 (November 26, 2019), https://doi.org/10.3389/fpsyg.2019.02589.

Andy Arnold, PhD, is an interoception researcher See https://www.andyjarnold.com.

I think that can be useful for social cognition in terms of the insula itself Further support for this statement has developed since the interview: Michael Datko et al., "Increased Insula Response to Interoceptive Attention Following Mindfulness Training Is Associated with Increased Body Trusting Among Patients with Depression," *Psychiatry Research Neuroimaging* vol. 327: 111559 (December 2022), https://doi.org/10.1016/j.pscychresns.2022.111559.

The insular cortex is activated when emotions are observed in others, including pain and aversive reactions to unpleasant foods Nadine Gogolla, "The Insular Cortex," *Current Biology* 27:12 (June 19, 2017): R580—R586, https://doi.org/10.1016/j.cub.2017.05.010.

characterized by feelings of warmth, concern, and care for the other, as well as a strong motivation to improve the other's well-being Tania Singer and Olga M. Klimecki, "Empathy and Compassion," *Current Biology* 24: 18 (September 22, 2014): R875–R878, https://doi.org/10.1016/j.cub.2014.06.054.

University of California, San Diego, neuroscientist Karen Dobkins, PhD, who lectures and teaches courses on connecting more mindfully with others See https://karendobkins.ucsd.edu/.

I saw that Karen had a public lecture on YouTube called "Compassion for the Perpetrator," which she'd released earlier in the year Karen Dobkins, "Compassion for the Perpetrator," recorded February 2020, posted September 12, 2020, https://youtu.be/bX6jrdAhArs.

Their work, which largely focuses on movement synchrony, has shown that behavioral synchrony among rowers is correlated with elevated pain thresholds B. Tarr, M. Slater, and E. Cohen, "Synchrony and Social Connection in Immersive Virtual Reality," *Scientific Reports 8, 3693 (February 27, 2018), https://doi.org/10.1038/s41598-018-21765-4.*

that synchrony and exertion during dance (including silent disco) independently raise pain thresholds and encourage social bonding Bronwyn Tarr et al., "Synchrony and Exertion During Dance Independently Raise Pain Threshold and Encourage Social Bonding," Biology Letters, vol 11, issue 10 (October 1, 2015), http://doi.org/10.1098/rsbl.2015.0767.

that movement synchrony forges social bonds across group divides Bahar Tunçgenç and Emma Cohen, "Movement Synchrony Forges Social Bonds across Group Divides," *Frontiers in Psychology* 7:782 (May 27, 2016), https://doi.org/10.3389/fpsyg.2016.00782.

that interpersonal movement synchrony facilitates pro-social behavior in children's peer-play Bahar Tunçgenç and Emma Cohen, "Interpersonal Movement Synchrony Facilitates Pro-Social Behavior in Children's Peer-Play," *Developmental Science* 21: 1 (December 18, 2016), https://doi.org/10.1111/desc.12505.

and that movement synchrony exclusively guides infants' social choices after 12 months of age Bahar Tunçgenç, Emma Cohen, and Christine Fawcett, "Rock with Me: The Role of Movement Synchrony in Infants' Social and Nonsocial Choices," *Child Development* 86: 3 (February 20, 2015): 976–84, https://doi.org/10.1111/cdev.12354.

The article acknowledges the complexity of defining and adopting mindfulness techniques and advocates for "attentional style" as a more useful and constructive concept Jonathan Gibson, "Mindfulness, Interoception, and the Body: A Contemporary Perspective," *Frontiers in Psychology* 10: 2012 (September 13, 2019), https://doi.org/10.3389/fpsyg.2019.02012.

"What is possible in Blackfoot may be impossible in English," says Leroy Little Bear, JD, a Blackfoot researcher and professor emeritus at the University of Lethbridge, in the book Blackfoot Physics F. David Peat, *Blackfoot Physics: A Journey into the Native American Universe* (Grand Rapids, MI: Phanes Press, 2002).

a verb that expressed the act of singing and included, as modifiers, one who sang and one who received the song Ibid.

without the sense of the invisible there can be no unity Marc Wittmann, "Modulations of the Experience of Self and Time," *Consciousness and Cognition* 38:172–81 (December 15, 2015), https://doi.org/10.1016/j.concog.2015.06.008.

"dyadic meditation"—where two people meditate together—may help us feel closer and more open with others Bethany E. Kok and Tania Singer, "Effects of Contemplative Dyads on Engagement and Perceived Social Connectedness Over 9 Months of Mental Training: A Randomized Clinical Trial," *JAMA Psychiatry* 74: 2 (February 1, 2017): 126–34, https://doi.org/10.1001/jamapsychiatry.2016.3360.

Similarly, research from the Virginia Commonwealth University School of Nursing found that the social connection aspect of yoga helps people manage depression Patricia Anne Kinser et al., "'A Feeling of Connectedness': Perspectives on a Gentle Yoga Intervention for Women with Major Depression," *Issues in Mental Health Nursing* 34: 6 (June 2013): 402–11, https://doi.org/10.3109/01612840.2012.762959.

the impulse to be more productive in any given sense arises from an internalized neoliberal ethos Deborah Lupton, *The Quantified Self: A Sociology of Self-Tracking* (Cambridge, UK: Polity, 2016).

Although there is limited research as of yet on Reiki and mental health, it is currently being offered at well-reputed medical institutions around the world, including the Cleveland Clinic in Ohio "Reiki," Cleveland Clinic, Wellness Institute Menu, accessed March 23, 2023, https://my.clevelandclinic.org/departments/wellness/integrative/treatments-services/reiki.

responsiveness to placebos, rather than a mere trick of the mind, can be traced to a complex series of measurable physiological reactions in the body; certain genetic makeups in patients even correlate with greater placebo response Gary Greenberg, "What If the Placebo Effect Isn't a Trick?" *New York Times*, November 7, 2018, https://www.nytimes.com/2018/11/07/magazine/placebo-effect-medicine.html.

I spoke to her about her research on MDMA and sociality Anya K. Bershad et al., "Effects of MDMA on Attention to Positive Social Cues and Pleasantness of Affective Touch," *Neuropsychopharmacology* 44: 10 (September 2019), 1698–1705, https://doi.org/10.1038/s41386-019-0402-z.

Several weeks later, I spoke with Gül Dölen, a neurobiologist at the Johns Hopkins Center for Psychedelics and Consciousness Research Parts of this interview appeared previously on the MIND Foundation website.

psychedelic drugs reopen a critical period in the brain when mice are sensitive to re-learning the reward value of social behaviors Romain Nardou et al., "Oxytocin-Dependent Reopening of a Social Reward Learning Critical Period with MDMA," *Nature* 569 (April 3, 2019): 116–20, https://doi.org/10.1038/s41586-019-1075-9.

the ones that started with a traumatic experience in childhood during this maximum sensitivity to the social environment, which happens during the social critical period An early stage in life when an organism is especially open to specific learning, emotional, or socializing experiences that occur as part of normal development and will not recur at a later stage. For example, the first three days of life are thought to constitute a critical period for imprinting in ducks, and there may be a critical period for language acquisition in human infants.

Part IV: Expanding the Possibility Space

the first and most powerful guiding light toward any understanding Audre Lorde, "Uses of the Erotic," *Sister Outsider: Essays and Speeches by Audre Lorde* (Berkeley: Crossing Press, 1984).

"Poetry Is Not a Luxury" Roxane Gay, ed., *The Selected Works of Audre Lorde* (New York: W. W. Norton & Company, 2020).

Esteban Kelly, a Philadelphia-based activist who co-founded the Anti-Oppression Resource & Training Alliance (AORTA), in a talk for the Barnard Center for Research on Women "What Is Transformative Justice?" Barnard Center for Research on Women, posted March 11, 2020, https://youtu.be/U-_BOFz5TXo.

"Consciousness," says the computational and cognitive neuroscientist Anil Seth, PhD, who helped popularize predictive coding, "has more to do with being alive than with being intelligent" Anil Seth, "We Are Beast Machines," *Nautilus*, December 18, 2021, https://nautil.us/we-are-beast-machines-2-238382/.

Philosopher and cognitive scientist Anna Ciaunica, PhD, takes this idea a step further, reminding us that we are alive within others before we even enter this world, in the womb of our mother Anna Ciaunica et al., "The First Prior: From Co-embodiment to Co-homeostasis in Early Life," *Consciousness and Cognition* 91 (May 2021): 103117, https://doi.org/10.1016/j.concog.2021.103117.

a convergence of "hard data and lived wisdom" From the Aeon + Psyche newsletter, "10 Years of Aeon," November 2022.

"Mystery abhors naked exposure and explanation. What chance does a rational justification have against the beauty and mystery of our lives?" John Daido Loori, *The Zen of Creativity: Cultivating Your Artistic Life* (New York: Ballantine Books, 2005).

While mushrooms are reproductive structures meant for spreading spores around, sclerotia are vegetative structures that serve no reproductive purpose Hamilton Morris, "Blood Spore," *Harper's*, July 2013, https://harpers.org/archive/2013/07/blood-spore/.

"The fact that their shared environment seems to be a stronger driver of gene content than shared ancestry," she observed, "is quite stunning" Jennifer Frazer, "Magic Mushroom Drug Evolved to Mess with Insect Brains," *Scientific American*, October 17, 2018, https://blogs.scientificamerican.com/artful-amoeba/magic-mushroom-drug-evolved-to-mess-with-insect-brains/.

This structure replacing the kernel isn't just general fungal material; it's a sclerotium Dieter Hagenbach and Lucius Werthmüller, *Mystic Chemist: The Life of Albert Hofmann and His Discovery of LSD* (Santa Fe: Synergetic Press, 2013).

It was used for medical purposes as early as the 16th century: for precipitating childbirth and stopping prenatal bleeding Aleksander Smakosz et al., "The Usage of Ergot (*Claviceps purpurea* [fr.] Tul.) in Obstetrics and Gynecology: A Historical Perspective," Toxins 13: 7 (July 15, 2021): 492, https://doi.org/10.3390/toxins13070492.

Although psychedelics can produce this effect, the researchers argue that "the capacity for pivotal mental states is an inherent property of the human brain itself" Ari Brouwer and Robin Lester Carhart-Harris, "Pivotal Mental States," *Journal of Psychopharmacology* 35: 4 (2021): 319–352, https://doi.org/10.1177/0269881120959637.

Practices

the leading edge of interoceptive dysregulation seen in depressive disorders Julie Dunne et al., "Losing Trust in Body Sensations: Interoceptive Awareness and Depression Symptom Severity among Primary Care Patients," *Journal of Affective Disorders* vol. 282 (2021): 1210–1219, https://doi.org/10.1016/j.jad.2020.12.092.

"Every mental event—that is, all states of consciousness—are said to ride the 'steed of wind' or 'energy' currents," write Norman Farb and colleagues Norman Farb et al., "Interoception, Contemplative Practice, and Health," *Frontiers in Psychology* 6 (June 9, 2015): 763, https://doi.org/10.3389/fpsyg.2015.00763.

appears to be a prerequisite for the other beneficial mechanisms to take place Gibson, "Mindfulness, Interoception, and the Body: A Contemporary Perspective."

Attentional switching (or flexible switching) is one of very few methods outlined in the current research literature Arnold, Winkielman, and Dobkins, "Interoception and Social Connection."

During a study on Mindfulness Based Stress Reduction, Carmody and Baer found that yoga was significantly associated with changes in mindfulness—specifically observing, acting with awareness, nonjudging, and nonreactivity as well as improvements in well-being, perceived stress levels, and several types of psychological symptoms Gibson, "Mindfulness, Interoception, and the Body: A Contemporary Perspective.".

a focus on sensation guided by the use of touch to support learning interoceptive awareness Cynthia J. Price and Carole Hooven, "Interoceptive Awareness Skills for Emotion Regulation: Theory and Approach of Mindful Awareness in Body-Oriented Therapy (MABT)," *Frontiers in Psychology* vol. 9: 798 (May 28, 2018), https://doi.org/10.3389/fpsyg.2018.00798.

a critical aspect of emotion regulation Ibid.

"Instead of facilitating emotion regulation via reappraisal or acceptance," writes study author and psychologist Erik C. Nook, PhD, "constructing an instance of a specific emotion category by giving it a name may 'crystalize' one's affective experience and make it more resistant to modification" Erik C. Nook, Ajay B. Satpute, and Kevin N. Ochsner, "Emotion Naming Impedes Both Cognitive

Reappraisal and Mindful Acceptance Strategies of Emotion Regulation," *Affective Science* 2 (2021): 187–98, https://doi.org/10.1007/s42761-021-00036-y.

"Respiration is unique compared to other sensations (such as the gastrointestinal one) insofar as conscious regulation can immediately impact respiratory processes, and respiratory processes can affect emotion and cognition," write Helen Weng, Jack Feldman, and colleagues in a recent article, "Interventions and Manipulations of Interoception" Helen Y. Weng et al., "Interventions and Manipulations of Interoception," *Trends in Neurosciences* 44: 1 (2021): 52–62, https://doi.org/10.1016/j.tins.2020.09.010.

ADIE is a novel therapy combining two modified heartbeat detection tasks with performance feedback and physical activity manipulation to transiently increase cardiac arousal Lisa Quadt et al., "Interoceptive Training to Target Anxiety in Autistic Adults (ADIE): A Single-Center, Superiority Randomized Controlled Trial," *eClinicalMedicine* 39 (August 1, 2021): 101042. https://doi.org/10.1016/j.eclinm.2021.101042.

has been shown to improve interoceptive processes and psychological well-being in several populations of study participants, including female adolescent ballet dancers Amie Wallman-Jones et al., "Feldenkrais to Improve Interoceptive Processes and Psychological Well-Being in Female Adolescent Ballet Dancers: A Feasibility Study," *Journal of Dance Education* (2022), https://doi.org/10.1080/15290824.2021.2009121.

Alexander technique instructor Imogen Ragone and trauma awareness activist Shay Seaborne discuss how awareness of bodily sensations can help us be more present, and explain how the technique improves interoception in this way Imogen Ragone and Shay Seaborne, "3 Gateways to Presence: Using Interoception, Proprioception, and Exteroception to Foster Resilience," August 30, 2020, in *Body Learning: The Alexander Technique*, podcast, https://podcasts.apple.com/us/podcast/3-gateways-to-presence-using-interoception-proprioception.

thus leading to resolution of the trauma response and the creation of new interoceptive experiences of agency and mastery Peter Payne, Peter A. Levine, and Mardi A. Crane-Godreau, "Somatic Experiencing: Using Interoception and Proprioception as Core Elements of Trauma Therapy," *Frontiers in Psychology* vol. 6 (February 4, 2015), https://doi.org/10.3389/fpsyg.2015.00093.

the creator of the Science of Social Justice framework for research and facilitation as well as the Systems-Based Awareness Map Mind Heart Consulting, accessed March 23, 2023, https://mindheartconsulting.com/.

engage in a compassionate viewing practice, noticing how images of different people on a screen elicit different bodily responses Dr. Sara King, "The Science of Social Justice," Wisdom 2.0, April 2022, https://youtu.be/wNg074CtEGk.

co-directing Embody Lab's Integrative Somatic Trauma Therapy Certification Program, and directing the Resilience Toolkit Certification Training Program "Facilitator Certification Program," The Resilience Toolkit, accessed March 23, 2023, https://theresiliencetoolkit.co/become-a-facilitator/.

Ndefo says in an interview with Lynn Fraser of the Kiloby Center for Recovery "Nkem Ndefo Resilience Toolkit," Kiloby Center for Recovery, September 8, 2022, https://youtu.be/Z7JbwrVTvAg.

what does it mean to live in a marginalized body? "Oppression and the Body," Penguin Random House, Specialty Retail, accessed March 23, 2023, https://www.penguinrandomhouseretail.com/book/?isbn=9781623172015.

Index